WILD PLANTS OF AMERICA

THE WILEY NATURE EDITIONS

AT THE WATER'S EDGE
Nature Study in Lakes, Streams, and Ponds,
by Alan M. Cvancara

WALKING THE WETLANDS
A Hiker's Guide to Common Plants and Animals of Marshes, Bogs, and Swamps,
by Janet Lyons and Sandra Jordan

MOUNTAINS
A Natural History and Hiking Guide,
by Margaret Fuller

THE OCEANS
A Book of Questions and Answers,
by Don Groves

WILD PLANTS OF AMERICA
A Select Guide for the Naturalist and Traveler,
by Richard M. Smith

NATURE NEARBY
An Outdoor Guide to America's 25 Most Visited Cities,
by Bill McMillon

Wild Plants of America

A Select Guide for the Naturalist and Traveler

RICHARD M. SMITH

Illustrations by the Author

WILEY

WILEY NATURE EDITIONS

John Wiley & Sons, Inc.

New York · Chichester · Brisbane · Toronto · Singapore

Library of Congress Cataloging-in-Publication Data

Smith, Richard M.
Wild plants of America.

(Wiley nature editions)
Includes index.
1. Botany—United States. 2. Plants—Identification.
I. Title. II. Series.
QK115 . S64 1989 581 . 973 89-8969
ISBN 0-471-51051-3
ISBN 0-471-62081-5 (pbk.)

Printed in the United States of America

89 90 10 9 8 7 6 5 4 3 2 1

TO JEANNE
who shared it all

Acknowledgments

For me, getting to know wild plants has always meant going to the places where they live. In my boyhood this was trudging up a steep trail peering for trailing arbutus, ambling through a sunny meadow aglow with goldenrods and asters, scuffling through fallen leaves to uncover the year's last gentian, crunching over crusted snow to tell the fortunes of swollen tree buds. This was the way my father taught me. It happened to be his only way, for I never knew him to uproot a plant from the wild for his garden, but I also saw it as the right way and am grateful for his example.

Many years later the desire to get out into the woods and fields coincided with the need for periodic respite from the confines of a New York office. Fortunately, I soon discovered the Torrey Botanical Club, and there followed a period of participation in their excellent field trips, which provided opportunities for plant hunting in unique areas that were within the reach of all city-dwellers but known to few. Encouraged by this, I was confident of finding a similar outlet when I moved to the botanical treasure house that is the southern Appalachians, and I was not disappointed. The Western Carolina Botanical Club proved to be an exceptional group of energetic, imaginative, and compatible individuals with interests identical to my own. Much that I learned from the members of these organizations has contributed to the writing of this book.

On many occasions in recent years I have been privileged to follow professional botanists as well as expert amateurs in the field, absorbing in the process a fund of knowledge that, to a much greater degree than I suspected at the time, has been found useful in preparing this work. Several friends suggested visits to areas I knew nothing about; others invited me to share the delights of botanizing in the natural surroundings of their homes. All across the country there are individuals who supplied me with information to facilitate my explorations and research, provided answers to technical questions, and extended other courtesies.

With assistance coming from so many quarters and taking so many forms, it would seem impossible to compile a list of those responsible without some inadvertent and embarrassing omissions. It is a risk worth taking, however, and I should like to record my appreciation to Cynthia Aulbach-Smith, John A. Bacone, Wade T. Batson, C. Ritchie Bell, Frank Bell, Sr., Clarence J. Bizet, Sylvia Bolton, Vickie Boultbee, John Reid Clonts, Caroline Douglass, Gussie Carrell Gray, Thomas E. Howard, August M. Kehr, Jeannie Wilson Kraus, Roy Lukes, Carol Mays, Ron Mezich, Charles F. Moore, Ken Moore, Will Murray, Laurie Osterndorf, Stephen Packard, Millie Pearson, Miles Peelle, Mildred Perkins, William D. Perkins, James D. Perry, Jesse Perry, Ruby Pharr, J. Dan Pittillo, Douglas Rayner, Connie and Bob Rhudy, Bruce Rowe, Paula Seamon, Thomas S. Shinn, Sr., Kathy and Craig Smith, Edwin F. Steffek, Jr., Arthur Stupka, Rob Sutter, George Wenzelburger, Judy Wick, B. Eugene Wofford, Thomas M. Woiwode, and Benny M. Woodham.

I am especially grateful to Dr. William A. Weber of the University of Colorado and Dr. William A. Niering of Connecticut College, who reviewed the entire manuscript, for their valued comments and suggestions.

I was fortunate in having had the assistance of three fine editors: Mary Kennan, who suggested the project and gave it her encouragement and support; David Sobel, my editor at Wiley, who never failed to temper expediency with patient understanding in the resolution of difficult problems; and Barbara Russiello, who expertly guided the book and its author from editing to production.

The gathering of material for this book has involved thousands of miles of travel and countless days on the trails. In virtually all of this I was accompanied and assisted by my wife Jeanne, who not only is an enthusiastic companion and expert plant spotter in the field, and a perceptive critic in the study, but has wholeheartedly and selflessly given her support to the project throughout its course. By any measure, it is her book as well.

Contents

Illustrations

CHAPTER 4

CHAPTER 5

CHAPTER 6

CHAPTER 7

CHAPTER 8

CHAPTER 9

CHAPTER 10

CHAPTER 11

CHAPTER 12

CHAPTER 13

CHAPTER 14

CHAPTER 15

CHAPTER 16

CHAPTER 17

CHAPTER 18

CHAPTER 19

CHAPTER 20

CHAPTER 21

CHAPTER 22

CHAPTER 23

CHAPTER 24

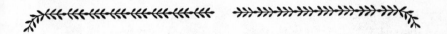

Introduction

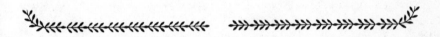

THE YEAR WAS 1584, and Queen Elizabeth I had sent for Sir Walter Raleigh. For decades the conquering forces of Spain had swept across the New World like wildfire, and now they were consolidating their hold upon Florida. The time had come, she decided, for England seriously to consider establishing a settlement in the territory that had been named Virginia in her honor. Accordingly, Raleigh dispatched two sea captains to reconnoiter the area, and on the 13th of July they dropped anchor near an island in present-day North Carolina. The Indians called it Roanoke, and it was to become the site of the first, albeit ill-fated, English colony in America.

Raleigh's charge included the obvious: There should be a safe harbor; a salubrious climate and fertile soil; tall trees for timber; fish, game, and wild plants for food; and native inhabitants who were amicable. But there was also the odd stipulation that they were to look for sassafras trees. Of all the exotic plants discovered by the Spaniards, this one had attracted the most intense interest. To the Indians it had some domestic uses, but before long the Europeans were crediting it with a multitude of imagined or exaggerated properties, and considered it a virtual panacea. As the English continued to settle Virginia, they sent whole shiploads of it home, and incredibly it became one of America's most coveted treasures.

Although the excitement over sassafras bordered on the irrational, the story is indicative of the appeal that plants previously unknown in the Old World would have for its botanists once exploration began in earnest at the start of the 18th century. At first the main focus was on the flora of the seacoast, especially near Charleston, where the benign winters were an advantage, but later the barrier of mountain wilderness to the west was breached. From other directions plant hunters probed the rest of the country, and soon the land was overrun by collectors seeking to satisfy patrons in many nations. John Bartram of

Philadelphia made innumerable shipments to his fellow Quaker, the London wool merchant Peter Collinson. André Michaux was dispatched by the French King Louis XVI to gather trees for Versailles, and the Scot John Fraser found himself in the service of Catherine the Great of Russia. Specimens were eagerly received by famous botanists like Linnaeus in Sweden and Gronovius in Holland.

Driven by an insatiable appetite for discovery, botanical explorers continued the search throughout the colonial period, the westward expansion, and the industrial age. Today this great abundance—or what is left of it—is accessible to us in a measure they could not have dreamed of.

Anyone who has the pleasant experience of crossing the United States, whether from coast to coast or from border to border, is certain to be impressed by the astonishing variety of its plant life. By the latter part of the 19th century it had become evident that in order for this bewildering diversity to be fully comprehended its components should somehow be sorted out and classified, and in 1889 this was accomplished in the Southwest by Dr. C. Hart Merriam, who created a systematic tabulation of what he called "life zones." Each of these zones possessed its unique combination of topographic, climatic, and other factors that fostered a distinctive group of plants and animals. Merriam's concept has since been modified by others, and some of these adaptations use labels that are more descriptive of vegetation types, such as alpine tundra and chaparral. Maps depicting these life zones—or the plant communities that distinguish them—look as though paints of many colors had dripped and been allowed to flow at random across the paper to form irregular blobs. Coniferous and

SASSAFRAS
Sassafras albidum

deciduous forests may be represented by different hues of green, grasslands by brown, deserts by patches of yellow, bits of tundra by isolated drops of blue, the Florida peninsula by pink to denote subtropical vegetation, and so on.

Just as an understanding of this zonation helped botanists to study the American flora, so does it provide us with a method of approach to what could otherwise be a frustrating experience as we move about over this vast and complex land. For this reason, the present text is arranged more according to types of habitat than on a strictly linear geographic plan. (Although the territory dealt with in this book is limited only by the boundaries of the conterminous United States, it was clear at the outset that because of space limitations not all of the vegetation zones could be included nor could every one of the states be represented.) It did seem, however, that the reader would appreciate being offered multiple possibilities for exploration within relatively small areas, rather than individual sites separated from each other by hundreds of miles. Accomplishing this was not much of a problem in the eastern part of the country, where the distances between examples of different natural communities tend to be short, but in the more spacious sections of the West it meant concentrating on several major national parks and omitting some botanically significant but isolated areas solely because of their remoteness.

In selecting trails for inclusion, botanical interest came first, but careful attention was also given to other criteria that it was felt would be important to many readers. Trails are accessible by car and negotiable on foot. Four-wheel-drive vehicles are not required, and there are no really strenuous or lengthy hikes (although some constitute the initial mile or so of a longer trail). In the few instances where water transportation is needed, public facilities are available; it is not necessary to own or charter a boat. All are open to the public, but policies governing seasonal closings, entrance fees, and so on will vary and should be ascertained by making inquiries in advance. Given the extreme variations in terrain—from deserts to mountains and seashores to swamps—there is always the possibility of encountering some of nature's less agreeable aspects, but the trails are safe and one need only take the reasonable precautions and exercise the prudent judgment that should be the rule whenever exploring the outdoors.

The places described in the body of this work are "natural" areas, in that the plants growing there either are native to the region or are aliens that have adapted so well to their new environment that they are now firmly established as a part of our flora. Their counterparts—places where man has purposely intervened—will be found in the Appendix, where a directory of selected botanic gardens and arboretums is furnished in the hope that it will be useful to readers whose inclinations or limitations may lead them in that direction. To some extent, these institutions were chosen on the basis of their having at least some space devoted to indigenous plants. Included are a number of sites where the species diversity has been exaggerated by the deliberate introduction of nonnatives. A few severely formal gardens and some that are primarily showcases for exotics also appear, simply because their reputations for excellence were felt to be sufficient reason for inclusion.

This is a firsthand account in that, with few exceptions, each of the sites described has been explored by the author and the plants attributed to it seen there at one time or another. Where it seemed advisable to do so because of elapsed time, return visits were made during preparation of the manuscript. In cases where even this was not possible, local authorities confirmed that the trails were open and the directions still valid.

The continued presence of plant species is somewhat harder to guarantee. Cataclysms like the New Madrid earthquakes of 1811–12, which created a new lake and changed the course of the Mississippi River, are now mere faded entries in our history, but we need only reflect upon the devastation of huge mangrove forests in the Everglades by Hurricane Donna in 1960, or the transformation of an entire mountain by the eruption of Mount St. Helens in 1980, to realize that comparable natural disasters could cause massive ruin even while these pages are being run through the presses.*

Much more likely to happen, of course, are less serious, short-range disturbances such as a dearth of rainfall inhibiting the desert's extravagant spring flowering, or a freak late frost literally nipping tree blossoms in the bud. Then there are the inexorable transitions brought about by natural vegetation changes. Also to be reckoned with is human intervention: although puny by comparison with the forces of nature, the power of humans is very effective when directed at habitat destruction, and more than one page has had to be discarded as its subject vanished before the bulldozer or the chain saw. Even more reprehensible is willful overcollection by commercial interests, which has resulted in the depletion of plant populations and even the extirpation of entire taxa.

Nowhere, of course, will you find anything approaching a full recital of all of the species that can be expected in a given area. For one thing, most trails are described as they appear at one particular time of year—usually, but not necessarily, the best season for finding wildflowers. Also, it was hoped that the exercise of conscious restraint would save some of the pleasures of discovery for the reader. Remember, too, that there may be other trails or sites nearby that would be worth looking into.

*This point was dramatically illustrated in 1988 when a wave of several thousand forest fires swept over the West. Whatever the degree of human culpability, this conflagration could not have occurred except as a result of a highly unusual combination of natural forces of unprecedented intensity and duration, including severe drought, high temperatures, and strong winds. Much of the public's attention focused on Yellowstone National Park, of which more than one-third was devastated. The chapter dealing with this area has, however, been left as it was originally written. Since it describes conditions as they had existed over many years of relative stability, it may prove useful during the period of regeneration in observing the kinds of changes that may be expected to occur periodically under the Park Service's "let burn" policy.

We are naturally curious about rare plants, and certainly education is near the top of the list of factors contributing to their preservation, but the question did arise as to how one could divulge specific information on their whereabouts in a widely distributed publication without running the risk of placing them in jeopardy. The solution, like the dilemma, has two parts: The first was to carefully consider which ones might be so vulnerable that increased traffic or publicity could prove to be deleterious, and omit them from the book. This decision was made easier by the knowledge that many of the organizations that have undertaken to protect such plants are willing to make suitable arrangements to accommodate responsible individuals. The second decision—which really applies to all of the information provided in the book—was quite simply to assume that readers will respect the spirit in which the information is imparted and use it conscientiously in ways that will foster rather than diminish this precious resource.

It is assumed that you will want to identify many of the plants as you come across them, and to do this a field guide should be considered essential. The most rudimentary ones contain photographs or drawings of selected flowers with descriptive captions, and offer little help beyond an opportunity to compare live specimens with printed pictures. Much to be preferred are those that provide some kind of key. Basically, a key is nothing more complicated than a series of paired questions about the plant, the answers to which, if correct, lead us down a succession of repeatedly forking paths to the name of the genus or species. By definition, field guides must be portable and therefore cannot be expected to list all species, so the best course is to select one whose geographic coverage corresponds rather closely to the region you are exploring. This can be supplemented by inexpensive guidebooks devoted to a single park or trail, which are available at many visitor centers or trailheads and, because of their narrow focus, are especially useful.

At the other end of the spectrum are the definitive manuals, which describe the flora of a specified region and usually include plants other than "wildflowers"—trees, grasses, ferns, and so forth. Of necessity they are bulky, and obviously they are not intended to be carried in the field. Although manuals are most often used by serious students and professional botanists, they can assist the amateur in solving problems of identification where field guides have failed, since they can be counted on to include *every* species that is to be found in the region covered.

Unfortunately, there is no single source of information concerning field guides for the traveler who wants to acquire them prior to departure. State universities that have herbaria (and native plant societies, which often are affiliated with such institutions) may hold the most promise. Contact might also be made with organizations involved with nature and conservation, such as the local chapters of the National Audubon Society and the Sierra Club, and the field offices of The Nature Conservancy. Many botanical gardens operate bookstores, as do the visitor centers at national parks, national monuments, and the like. Private nonprofit natural history associations provide books and other interpretive materials for the National Park Service, and accept mail orders. Sellers of out-of-print and hard-to-find titles are another possible source; their catalogs will make it clear which books are field guides. State agencies concerned with tourism and natural resources can help with highway maps, directories of state parks, and the like, but you should not expect much with specific reference to botany. Finally, when you arrive at your destination be sure to check out the local stores for books of regional interest that may have had only limited distribution.

The subject of identification can hardly be left without at least a brief mention of botanical nomenclature. The common names of plants have become convenient through long usage, and are the ones we use conversationally when we are sure that people will know what we are talking about. The trouble is that a common name loses its usefulness when, as has frequently happened, the same name has been given to several different species or multiple names have been conferred on a single species as a result of regional preferences. For this reason, positive identification of a plant often requires that we ascertain its scientific name.

There are occasions, however, when new information suggests that even a long-established scientific name should be vacated in favor of one that has precedence under the rules of nomenclature. This creates a situation where one may encounter different names in the literature without realizing that they refer to the same plant. In order to best serve the majority of users of this book, it was decided to use the names that probably will match the ones they will find in the current popular field guides. This, on the other hand, should cause no great inconvenience for those whose expertise is at or near the professional level, since they will be quite familiar with the synonymy.

There is no question that the text would read more smoothly without the intrusion of scientific names, but to omit this important, and often necessary, information would have greatly diminished the book's utility. As a compromise, common names are used throughout but with the botanical equivalent for each species added parenthetically where it first appears in a chapter.

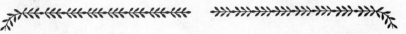

1

High Mountains of New England

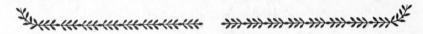

IF WE USE the term "high" in an ecological sense to denote mountains that rise to such inhospitable elevations that no trees are able to grow on their summits, we find that there are only three areas in New England where they meet the definition: Mount Katahdin in Maine, the White Mountains of New Hampshire, and the Green Mountains of Vermont. In fact, the only other peak in the entire Northeast that would qualify is Mount Marcy, the highest point in New York State. (No mountains in the Southeast have a tree limit, but this has nothing to do with altitude; many of them are even taller than those in the Northeast. Rather, it is a function of the milder climatic conditions found in the southern latitudes.)

The area lying above treeline—the alpine zone—has a fascination all its own, largely due to a unique and beautiful flora; and, as we shall see, this is no less true in the Northeast than in the western ranges where its counterparts are numerous. But let us first turn to the rich, cool forests that clothe the slopes and fringe the ponds in these ancient northeastern mountains.

Immediately below timberline, the boreal plant community is exemplified by a number of coniferous species. Typically northern is Balsam Fir *(Abies balsamea),* which is grown extensively and marketed as a Christmas tree, often labeled Nova Scotia Balsam. White, Red, and Black Spruces *(Picea glauca, P. rubens,* and *P. mariana)* are common. The last is found almost exclusively in or near sphagnum bogs, as is the deciduous Tamarack or American Larch *(Larix laricina).* Another moisture-loving tree is Northern White Cedar *(Thuja occidentalis),* which is frequently used in landscaping; the emerald green fanlike sprays of flattened leaves and its alternate name Arbor Vitae are familiar to many.

Numerous broad-leaved tree species are components of both this boreal forest and the "transition" community that occupies the next lower altitudinal

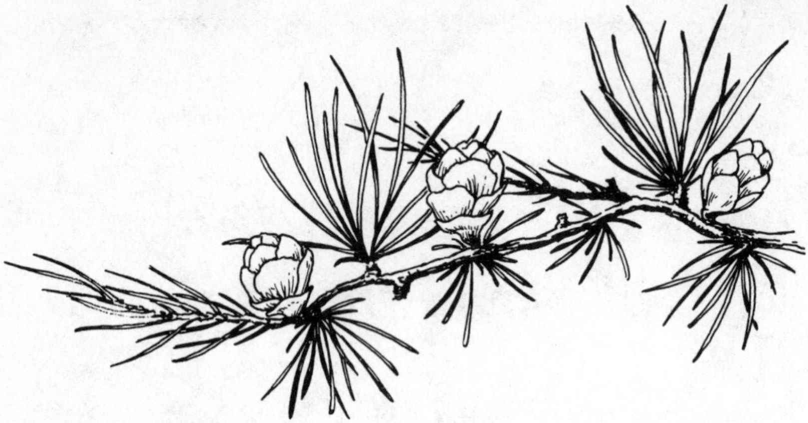

TAMARACK
Larix laricina

zone. Their presence is strikingly evident during the fall foliage season. The luminous orange of the Sugar Maples *(Acer saccharum)* and the variegated red-to-yellow hues of Red Maple *(A. rubrum)* are punctuated by chalk white, golden-crowned Paper Birches *(Betula papyrifera),* while the scarlet fruit clusters of American Mountain Ash *(Sorbus americana)* provide glowing accents. Add to this the contrasting greens of the conifers and you have a spectacle that would be difficult to match anywhere.

THE PINE TREE STATE

Just seeing or hearing the words "Maine Woods" is enough to create a vivid picture in our imagination, even though we may never have been Down East. We can almost see the gleam of white birches against deep green shadows, hear the soft lapping of lakeshore waves and the rustling of poplar leaves muffled by carpets of moss, and breathe in the cool, moist air redolent of pine and fir. Perhaps our senses have been helped in forming this preconception by some earlier associations, vague and nearly forgotten: the gift of a little sachet pillow stuffed with "balsam" needles, the smooth feel of a toy birchbark canoe, glances at a scenic hunting and fishing calendar on the wall.

Then, when we finally visit Maine, we make the happy discovery that the image was an accurate one. The sights, sounds, and smells are all there, and if we are surprised at anything it is at finding the forests to be so extensive. From the borders of Canada and New Hampshire, wherever humans have permitted, forests spread across the land all the way down to the seacoast, stopping only where gnarled rocky fingers dip into the cold waters of the Atlantic.

Mount Katahdin and Baxter State Park

Mount Katahdin is the highest point in Maine, falling short by just a dozen feet of reaching a mile above sea level. How much higher it was originally, and what its shape might have been, can hardly be guessed. It may have risen sharply to a peak like those in the Grand Tetons until, along with many other northeastern mountains, it was sheared and gouged by glaciers. Now it is only in certain aspects that we can discern vestiges, in low profile, of an old pyramidal form.

Katahdin is the centerpiece of the 200,000-acre Baxter State Park, a gift to the people of Maine from Percival P. Baxter, a former governor of exceptional foresight, determination, and generosity. Frustrated in his efforts to convince the legislature that it should rescue this wilderness area from threatened exploitation and speculation, he quietly began a 30-year program of buying it up himself, parcel by parcel, and turning it over to the public. Katahdin itself has a special significance as the northern terminus of the Appalachian Trail, which extends for more than 2,000 miles through 14 states to Springer Mountain in Georgia.

On highway maps Baxter appears as a neat rectangle in north-central Maine and could not be easier to find, but it also will be evident that it is relatively remote. Driving along its principal roads can be thoroughly enjoyable and afford excellent viewing of plant life, and beautiful trails are accessible for day hikes. It would be a mistake, though, to assume that all parts of the interior are suited to casual walking, for simple logistics make backpacking a necessity in order to reach much of the region.

This can be a source of regret because, for one thing, Katahdin is still lofty enough despite eons of erosion to possess an alpine environment on its summit. However, the shortest adequately marked trail to the top from a drivable road involves a 5-mile ascent, a vertical climb of more than 4,100 feet, a rough footway, and hazardous, strenuous clambering over large boulders. (Fortunately, though, the Presidential Range in neighboring New Hampshire has alpine areas that not only are more easily reached but are greater in extent, and provide the best opportunities east of the Rockies to examine this unusual biome.)

For those traveling by car and wanting to enjoy short walks, a good selection of trails can be found in the Deadwater Mountains (the name was bestowed by Henry David Thoreau) in the upper part of Baxter State Park. Access is from Grand Lake Road, which crosses the park from Grand Lake Matagamon at the northeast corner to Nesowadnehunk Lake on the western border, and the following directions assume this to be the direction of travel.

The first reference point is the site of Trout Brook Farm, where there is a small campground. Large cleared areas dating back to 1837 when lumbering camps were set up here have provided space for impressive displays of Canada Lilies *(Lilium canadense),* which are near the place where Thoreau recorded in his journal having collected the bulbs for soup. The cleared areas have also provided opportunities for many alien species to establish themselves. Of

BEAKED HAZEL
Corylus cornuta

particular interest is Silvery Cinquefoil *(Potentilla argentea)*, so named for the white, woolly hairs that cover the undersides of the leaves.

In summer the roadside ditches should be watched for Purple Fringed Orchids *(Platanthera psycodes)*. On the edges of the woods, and far less conspicuous, are Beaked Hazels *(Corylus cornuta)*; in this species the bracts coalesce around the fruit and become greatly prolonged.

A short distance west of Trout Brook Farm, Fowler Brook suddenly appears on the left, a pretty little placid stream meandering among tufts of sedges and grasses. Here is the beginning of the attractive Fowler Pond Trail, which after passing through open woods comes out upon Lower Fowler Pond. At its outlet the sun-bleached wreckage of downed trees has piled up, and in the resulting accumulation of detritus a disparate assortment of herbaceous plants has taken hold: Cattails *(Typha latifolia)* and Water Parsnip *(Sium suave)* in shallow water, Large Canada St. John's-wort *(Hypericum majus)* in muddy gravel, and Round-leaved Sundew *(Drosera rotundifolia)* in tiny pockets of damp soil.

Depending upon the time one has available, it is possible to skirt the shores of both Lower and Middle Fowler Ponds (they are only a half mile apart), passing through White Pine, White Birch, and Northern White Cedar in turn, with Swamp Roses *(Rosa palustris)* occasionally brightening the water's edge. The fact that the trail is not far from the road seems to mean nothing as far as wildlife is concerned, and you should feel cheated if you do not see deer at least once in the course of a few hours' walking.

Returning to Grand Lake Road and continuing westward, you can reach another trail from South Branch Pond Campground, which is at the end of a short spur on the left. Roadside plants on the way may include Red-berried Elder *(Sambucus pubens),* Bristly Sarsaparilla *(Aralia hispida),* and Spreading Dogbane *(Apocynum androsaemifolium).* Several mints, including Woundwort *(Stachys palustris)* and Mad-dog and Marsh Skullcaps *(Scutellaria lateriflora* and *S. galericulata),* have been observed.

The campground at Lower South Branch Pond, which has facilities for swimming and boating, is the trailhead for hikes to Upper South Branch Pond, and beyond. Among the woodland plants along this trail are Shinleaf *(Pyrola elliptica),* Red Baneberry *(Actaea rubra)*—some with white fruits—and Spikenard *(Aralia racemosa),* while Spatterdock *(Nuphar variegatum)* is profuse in backwater areas. One of the more interesting ferns is Rusty Woodsia *(Woodsia ilvensis).*

On the way out, midway along the spur road, there is a short path to a waterfall with carpets of Bunchberry *(Cornus canadensis).* To a southerner this could suggest a sprig from a Flowering Dogwood tree fallen to earth, its leaves rearranged in a whorl and the floral bracts drawn to a point. Also here is the delicate Twinflower *(Linnaea borealis),* a favorite of the great Swedish botanist Linnaeus and the flower he chose to bear his name.

Dwelly Pond, about 12 miles farther west on the right side of Grand Lake Road, is a special place. Here the shore is covered with typical bog and

BUNCHBERRY
Cornus canadensis

LABRADOR TEA
Ledum groenlandicum

lakeshore shrubs: Leatherleaf *(Chamaedaphne calyculata),* Labrador Tea *(Ledum groenlandicum),* Bog Rosemary *(Andromeda glaucophylla),* Sheep Laurel *(Kalmia angustifolia),* and Buttonbush *(Cephalanthus occidentalis).* In small openings, and in the adjacent woods, are Goldthread *(Coptis groenlandica),* Wood Sorrel *(Oxalis montana),* Three-leaved False Solomon's-seal *(Smilacina trifolia),* Clayton's Bedstraw *(Galium tinctorium),* and the insectivorous Pitcher Plant *(Sarracenia purpurea).*

For several years the main attraction at Dwelly Pond was a Bull moose that came regularly to feed on aquatic vegetation not far from shore, seeming to materialize out of nowhere at half-past four every afternoon. For at least one observer, the short wait for the curtain to go up on this show was enlivened by a goldeneye parading her little flotilla of ducklings nearby—a prologue that for a few days was as dependable as the main event. Although moose are not usually as predictable as this one, an alert motorist has a good chance of spotting these huge animals just about anywhere along this road.

THE WHITE MOUNTAINS

If you have ever seen the White Mountains in winter, wearing a mantle of glistening new snow against a sapphire sky, you would wonder what other name they could possibly have been given. Even at other seasons, sunlight shimmering on the barren masses of mica schist on the high peaks gives them an aspect very different from that of mountains that are sheathed in trees all the way to the top.

New Hampshire—the most mountainous of all the New England states—is justifiably proud of this magnificent range, which stretches from Vermont to Maine and is largely embraced by the White Mountain National Forest. Mount Washington, at 6,288 feet, is the highest mountain in the northeastern United States and is said to have been sighted from sea as early as 1605. A giant among giants, it is the keystone of the Presidential Range, an elite group that includes Mount Adams, Mount Jefferson, and Mount Monroe. These and other peaks are connected by the unique Appalachian Mountain Club hut system, a series of eight lodges located a day's hike apart.

In 1861 a carriage road was built up the eastern flank of Mount Washington, and today its successor, the Mt. Washington Auto Road, offers the easiest, most comfortable means of attaining the summit. An interesting alternative is to take the coal-burning trains of the Mt. Washington Cog Railway up the western side; these leave from the Base Station, which is a 4-mile drive from Fabyan on U.S. 302.

There are, of course, several routes by which the summit can be reached on foot, but it is a long way up there and foot travel over this distance does not allow much time for studying plant life. (Presumably hikers will put this book down and consult trail guides instead.) This is not to imply that there are not trails of more modest length nearby; there are, of course, and some of these will be described later.

The Arctic-Alpine Zone

For the moment, let us assume that you are primarily interested in seeing the alpine flora on the backbone of the Presidential Range, and that to save time or for other reasons you have elected not to walk to the top. The Auto Road enables you to drive—or be driven, if you prefer—the 8 miles from U.S. 16 at Glen House to the summit of Mount Washington. The total gain in elevation is 4,700 feet, but switchbacks keep the average grade down to 12 percent. There are spectacular views along the second half (treeline is passed about halfway up) and from the summit, where on those admittedly rare cloudless days you can see for 65 miles in every direction.

The landscape on top is a jumble of squarish boulders set at every conceivable angle. Seascape might be a more apt term, for the technical name for it is *felsenmeer,* which translates to "sea of rocks." Long ago, as attested to by their encrustation of lichens, these blocks of granite were broken and heaved by the powerful force of water that had been alternately subjected to freezing and thawing temperatures. Such extreme variations in climate, all too evident to hikers even today, are due to the exposed position of Mount Washington at a point where violent storm systems collide with a fury unparalleled anywhere except possibly near the poles.

Here and there a thin line of bare soil indicates a foot trail, marked at frequent intervals by cairns built of small rocks. One of these—the Appalachian Trail—drops away toward the southwest for a mile and a half into the col between Mount Washington and Mount Monroe. Down there the sun picks up what looks like two tiny bits of broken mirror: the Lakes of the Clouds. Next to them is a structure that seems small at this distance, but actually it is the largest of the Appalachian Mountain Club huts, with a capacity for housing 90 guests.

About a mile below the summit, the Alpine Garden Trail takes off from the Auto Road in the direction of Tuckerman Ravine, and this offers excellent opportunities for examining the tundra flora. Once you start walking you realize what an abundance of vegetation enlivens this apparent wasteland; the total number of species exceeds one hundred. The most prevalent ones are predominantly green: Sedges (*Carex* spp.), Rushes (*Juncus* spp.), Woodrushes (*Luzula* spp.), and several genera of Grasses. Occasionally these manage to grow in sizable drifts, which are called "lawns." Of the nonvascular plants, there are cushions of Haircap Moss *(Polytrichum juniperinum)* and clumps of white filigreelike Reindeer Lichen (*Cladonia* spp.).

The showier flowering plants are low in stature and huddle beneath or between rocks. Here they are sheltered from the winds, absorb radiated warmth, and conserve precious moisture while they compress flowering and fruiting into an extremely short season.

Heath shrubs are well represented, which is not surprising given the cold environment. Among the more common ones are Bog Bilberry *(Vaccinium uliginosum),* Mountain Cranberry *(V. vitis-idaea)* with four-lobed blossoms and large red berries, and the deciduous Alpine Bearberry *(Arctostaphylos alpina).* A tiny rhododendron known as Lapland Rosebay *(Rhododendron lapponicum)* bursts into bloom early with purple half-inch flowers; Alpine Azalea *(Loiseleuria procumbens),* a very durable little shrub that hugs the ground tightly, has pink blossoms only half that size. Moss Plant *(Cassiope hypnoides)* and Mountain Heath *(Phyllodoce caerulea)* each have bell-shaped flowers, which are white and pink, respectively; the first derives its common and specific names from the multitude of tiny leaves covering its stems.

These dwarf shrub communities have representatives of other families as well, notably Diapensia *(D. lapponica),* an attractive circumpolar species. Most of the other plants found here also have a generally northern distribution, from the ordinary-looking Alpine Goldenrod *(Solidago cutleri)* to the Dwarf Yellow Cinquefoil *(Potentilla robbinsiana),* an extremely rare and endangered species that is endemic to the high peaks of New England.

A number of others are also encountered well to the south. The Alpine Bluet is merely a white form of the familiar *Houstonia caerulea,* and the large-flowered Mt. Washington Avens *(Geum peckii)* is so similar to the *G. radiatum* of the southern Appalachians that the latter was once considered a variety of it. The cheerful Greenland Sandwort *(Arenaria groenlandica),* which is almost ubiquitous in this alpine habitat, recurs at high altitudes southward, as does the white-flowered Wine-leaved Cinquefoil *(Potentilla tridentata).*

Other Mount Washington Trails

Two popular trails, roughly similar and both easily accessible from U.S. 16, offer a choice for hiking a mile or two up the lower slopes of Mount Washington. One is the Tuckerman Ravine Trail from the Appalachian Mountain Club's Pinkham Notch Camp; if you go as far as Hermit Lake, the vertical climb is 1,900 feet. The other is the Glen Boulder Trail, named for an enormous glacial erratic that can be seen perched on the slope about 1,700 feet above the level of the highway.

Both feature interesting waterfalls. Crystal Cascades is along the Tuckerman Ravine Trail just a quarter mile from the start. Glen Ellis Falls is on the opposite side of the road from the Glen Boulder trailhead, but is well worth the short walk as it is one of the most beautiful waterfalls in the White Mountains.

Taking off at an elevation of about 2,000 feet, the trails pass through mixed hardwood forests with such trees as Yellow Birch *(Betula lutea)* and Pin or Fire Cherry *(Prunus pensylvanica),* but these eventually give way to Spruces and Firs. The handsome Paper Birch, always a welcome sight, is one tree that seems oblivious to changes in vegetation zones and keeps climbing up and up until there are no trees of any kind to keep it company.

Timberline, defined as the level above which normal trees do not grow, is reached quite abruptly. Dwarfed Willows (*Salix* spp.) and Mountain Alders *(Alnus crispa)* appear, and the conifers become grotesquely stunted and contorted. Spruces and Firs, together with some Birches and Mountain Ash, lie prostrate forming a tangled, springy mat of needled branches, sometimes only inches thick but dense enough to walk on. This is the beginning of *krummholz* (which means "crooked tree"), and although it may be a sunny, almost windless day, you know that the low scrubby growth pattern of these trees is their way of dealing with the fierce winds that whip across the land most of the time, threatening them with desiccation and slashing the tender new shoots with sharp crystals of sleet and snow. These are mature trees, and many have root systems that are more than a century old. They frequently spread horizontally by means of layering; heavy snows weigh down the lowest branches into contact with the soil, whereupon they take root and start a new generation of satellite trees.

As you ascend, the woody growth is reduced in extent and some of the larger flowering plants brighten the landscape, among them Fireweed *(Epilobium angustifolium)* and Pearly Everlasting *(Anaphalis margaritacea).* The bigger rocks are adorned with crustose lichens in yellow and orange as well as less vivid hues.

Ultimately the *krummholz* will phase out, to be replaced by the scrambled mass of boulders with their cushions of "belly plants." But on this comparatively well-protected leeward side of the mountain, that will not occur until an altitude of 5,000 feet or more has been reached.

2

Rocky Coasts and Sandy Shores

CAPE ANN, the little knob of land jutting out into the Atlantic above Boston, seems made for artists. The Gloucester fishing fleet is as picturesque as it is famous, the towns have stately white houses and steepled churches amid tall elms, and the country lanes are bordered by rose-covered cottages weathered to a silvery gray by the salt-laden breezes. On a stone wharf in Rockport harbor sits the little red shack known to painters the world over as Motif No. 1.

Nowhere is there monotony, and certainly this is true of the coastline, as exemplified by one particular crescent-shaped pocket beach of pure glistening sand flanked by dark, rocky promontories. I remember thinking that if you looked straight ahead you could imagine being on the smooth sands of Cape Cod, but that merely by turning your head you could be transported to the craggy coast of Maine. The reality is that this little cove is in a transition zone, situated between two very different kinds of shoreline and combining elements of each.

DOWN EAST

The coast of Maine from the Canadian border down to Portland has a configuration quite unlike any other along our Atlantic shores. Maps show clearly that it has none of the smooth, curving lines that characterize the seaward contours of other states to the south, but instead is deeply incised by innumerable bays and estuaries and fringed by a multitude of peninsulas. Someone has figured out that if this erose coastline—which measures less than 200 miles from point to point—could be drawn out into a straight line without any of these kinks and squiggles, it would reach all the way to South America!

Reasons for its peculiar form can be found in its origins, for here we have the kind of seacoast that is caused largely by submergence. The weight exerted by the massive glaciers greatly depressed what was then a broad coastal plain. As meltwater from the warming ice masses raised the level of the sea, this vast area was inundated and the waters gradually worked their way up into the valleys to form an irregular, jagged shoreline. Some portions, elevated but not high enough to rise above the surface, are the now-celebrated offshore fishing banks; a few emersed mountaintops are visible today as islands. (This is, of course, a much simplified explanation, but a detailed recounting of the successive and alternating periods of flooding and uplifting would belong more to the field of geology.)

This rockbound coast provides a stable support for marine macroalgae, or seaweeds as we commonly call them. Extending to the very mouth of the Bay of Fundy, this coast is noted for some of the world's greatest tides, and the exceptionally wide intertidal zone affords superb opportunities for becoming closely acquainted with these plants. Most of the large kelps will still elude us as they must keep to deep water, as will many of the delicate, light-sensitive red algae, but there are exceptions. The brown, leathery Horsetail Kelp *(Laminaria digitata)* may occasionally be exposed at low tide, and two reddish species, both of which have found a place in the human diet, can be seen in the lower tide pools: the lobed straplike fronds of Dulse *(Rhodymenia palmata)* and the multibranched tufts of the so-called Irish Moss *(Chondrus crispus)*.

Higher up are the Rockweeds, or Wracks. These will be represented by *Ascophyllum nodosum* or by one of several species of *Fucus,* depending upon the force of the surf with which it must contend in that particular place. When immersed, their fronds are buoyed up by air-filled bladders, but when the tide is out they lie strewn about in a dense, tangled, rubbery mass. The dividing line

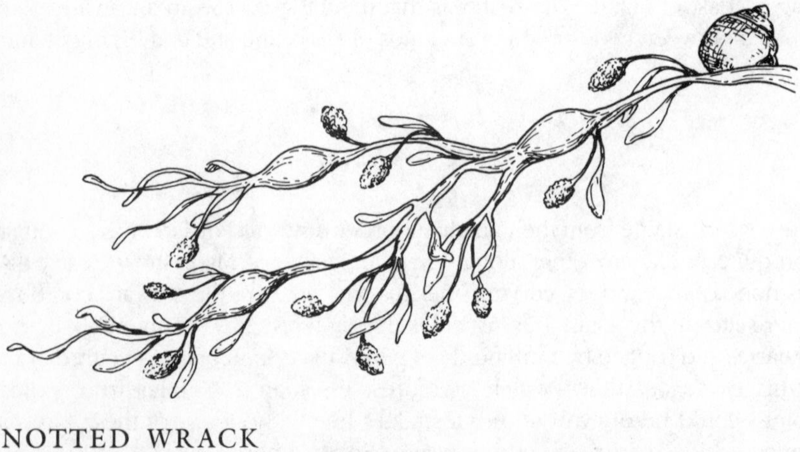

KNOTTED WRACK
Ascophyllum nodosum

between sea and land, where the rock is lapped briefly by the highest tides, is marked by a band of gelatinous black stain. This so-called black zone is composed of the most primitive of all plants, blue-green algae, chiefly species of *Calothrix*.

Beyond the reach of even the highest spring tides but within the range of wind-driven salt spray, plant life is limited almost entirely to lichens. One of the most conspicuous is the crustose *Xanthoria parietina,* which paints the rock surfaces with splotches of bright yellow and orange. Tide pools in the spray zone are made brackish by infusions of fresh water, and this is conducive to the growth of green filamentous algae such as *Enteromorpha intestinalis* and *Cladophora* spp.

Acadia National Park

To the majority of Maine vacationers, who choose to visit the scenic coast in preference to the more remote north woods, it is a source of gratification that an easily accessible national park has been provided just offshore.

Most of Acadia National Park is located on Mount Desert Island, southeast of Bangor. The names reveal its early French history. Once part of the vast region known as L'Acadie, the bare-topped island was called "L'Isle des Monts Deserts" by its discoverer, Samuel de Champlain, in 1604, and its highest point, Cadillac Mountain, takes its name from Antoine de la Mothe Cadillac, who received the island as a land grant from the Sun King, Louis XIV, and later founded the city of Detroit.

Eventually Mount Desert Island passed through British and into American hands, and by the late 19th century the quiet village of Bar Harbor had become a favorite summer resort of the country's millionaires. Fortunately, however, both residents and visitors began donating land to the government in the 1900s, and these early efforts at conservation culminated in the establishment of a unique national park aggregating 65 square miles in extent.

To reach the Mount Desert Island portion of the park, take S.R. 3 and stop at the Hulls Cove Visitor Center. This marks the beginning of a 20-mile loop road with branches leading to various natural features, one being a road to the top of Cadillac Mountain. There are more than a hundred miles of trails, varying from level to steep, as well as a system of gravel carriage paths built by John D. Rockefeller, Jr., which now provide automobile-free routes for bicycling and horseback riding as well as hiking.

As you follow the eastern shoreline on the loop road, you will see evidence of the disastrous 1947 fire, which burned for 26 days over 77,000 acres and wiped out most of Bar Harbor's remaining mansions. The voids in the island's covering of Spruce and Fir left by this conflagration were quickly invaded by such trees as Quaking and Bigtooth Aspens (*Populus tremuloides* and *P. grandidentata*), Paper and Gray Birches (*Betula papyrifera* and *B. populifolia*), and

FIREWEED
Epilobium angustifolium

Fire Cherry *(Prunus pensylvanica);* by carpets of Low Blueberry *(Vaccinium angustifolium);* and, most spectacularly, by enormous drifts of Pink Fireweed *(Epilobium angustifolium)* punctuated by the white of Pearly Everlasting *(Anaphalis margaritacea).*

About 6 miles from the start is Sieur de Monts Spring, where there is a Nature Center, a Museum, and the Wild Gardens of Acadia. This last is a collection of labeled plants from various areas of the park, including its bogs, fields, and shores, and will be especially helpful to anyone unfamiliar with the flora of the northeast coast. A short detour onto S.R. 3 South takes you to The Tarn, an attractive lake with masses of showy Water Lilies *(Nymphaea odorata),* Spatterdock *(Nuphar variegatum),* and Pickerel Weed *(Pontederia cordata)* and the tiny but very numerous Common Bladderwort *(Utricularia vulgaris).*

Returning to the loop road and continuing for a few miles, you come to Sand Beach. Beaches are rare in this part of Maine, and here the sand contains a surprisingly high percentage of pulverized shells of marine organisms. Of the several walks that can be made from here, the easiest and perhaps the most interesting is the 1.8-mile (each way) Shore Path, which follows the water's edge past the violently pounding surf at Thunder Hole to Otter Point. From this trail you can discern the extent of the 1947 fire by noting where the small, young conifers with their accompaniment of Bigtooth Aspens, Birches, and Fireweed are replaced by older Pitch Pines *(Pinus rigida)* and groves of tall White Spruces

(Picea glauca) that escaped being burned. Among the more robust flowering plants near the shore are Red-berried Elder *(Sambucus pubens),* Salt Spray Rose *(Rosa rugosa),* Seaside Goldenrod *(Solidago sempervirens),* Common Tansy *(Tanacetum vulgare),* and Meadowsweet *(Spiraea latifolia).* On rock ledges we find the creeping Northern Mountain Cranberry *(Vaccinium vitis-idaea)* and spreading mats of Black Crowberry *(Empetrum nigrum),* along with the delicate Round-leaved Harebell *(Campanula rotundifolia).*

After a while the road turns inland to Jordan Pond, where there is access to carriage roads and a half-mile self-guided nature path. This trail, with its explanatory booklet, provides an excellent introduction to northern trees like White and Red Spruce, Balsam Fir *(Abies balsamea),* Northern White Cedar *(Thuja occidentalis),* Striped Maple *(Acer pensylvanicum),* and American Beech *(Fagus grandifolia).* The pamphlet also calls attention to several fascinating lichens: Pyxie Cups *(Cladonia pyxidata),* scarlet-tipped British Soldiers *(C. cristatella),* Lungwort *(Lobaria pulmonaria),* and Old Man's Beard (*Usnea* spp.). A longer trail encircles Jordan Pond for a distance of 3.3 miles. This gets you farther away from the influences of civilization and thus is more rewarding, but there is some scrambling over loose morainal rock.

The climax of any trip to Acadia National Park is sure to be the drive up the spur road to the top of Cadillac Mountain. Standing on this granite eminence, you are at the highest point on the U.S. Atlantic coast, 1,530 feet above the sea. You can look down on the Porcupine Islands in Frenchman Bay, and from a mile-long foot trail around the summit you can see many of the places you explored earlier from the loop road. If you are here around the first of June you will be in time for the beautiful pink Rhodora *(Rhododendron canadense).* Later in the year, look for the strange Yellow Rattle *(Rhinanthus crista-galli)* and the dainty white flowers of the closely related Eyebright (*Euphrasia* spp.).

Although it is tempting to devote all of one's time to the attractions that are concentrated along the loop road, a trip down the western half of the island should be made if at all possible. By taking S.R. 102 and 102A beyond the Seawall Campground, you arrive at a bog known as Big Heath. Here you will see blooming, in their respective seasons, Arethusa *(Arethusa bulbosa)* and Grass Pink Orchids *(Calopogon tuberosus),* carnivorous Pitcher Plants *(Sarracenia purpurea)* and Sundews *(Drosera rotundifolia),* and a panoply of ericaceous plants including Rhodora, Sweet Gale *(Myrica gale),* Labrador Tea *(Ledum groenlandicum),* Bog Laurel *(Kalmia polifolia),* and Bog Rosemary *(Andromeda glaucophylla).* At Ship Harbor, an interpretive nature trail presents interesting facts concerning the history and ecology of this low-lying area.

A small section of Acadia National Park occupies part of Isle au Haut (reached by mailboat from Stonington). The third component—the only part on the mainland—is at Schoodic Point, a promontory on the opposite side of Frenchman Bay.

A one-way road loops around the tip of Schoodic Peninsula from the park entrance south of Winter Harbor, rejoining S.R. 186 at Birch Harbor. This follows

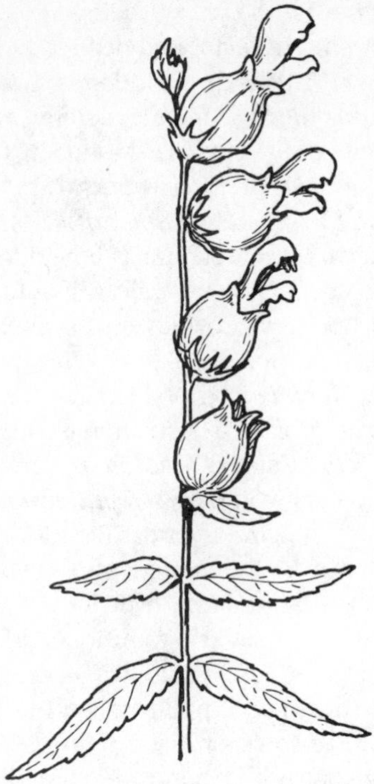

YELLOW RATTLE
Rhinanthus crista-galli

the rugged coastline through Jack Pines *(Pinus banksiana).* There are numerous turnouts providing access to the shore, and trails to Schoodic Head and the Anvil, the peninsula's high points. Much of the plant life is that adapted to sparse soil conditions in exposed situations: Wine-leaved Cinquefoil *(Potentilla tridentata),* Crowberry, Mountain Cranberry, and Eyebright.

Wolf Neck Woods

No less than 16 of Maine's state parks are strung out along its coastline from New Hampshire to New Brunswick. A fine example, made eminently accessible by its location a mere 20 miles from the largest city, Portland, is Wolf Neck Woods State Park. It is reached from the L.L. Bean store in Freeport by turning east on Bow Street for 1.4 miles, right at the fork for another 0.9 mile, then right again on Wolf Neck Road.

A network of trails crisscrosses the peninsula between the Harraseeket River estuary and the salt waters of Casco Bay, and leads through an extraordinary assortment of ecosystems. There is, for example, a woodland rich in wildflowers such as Trailing Arbutus *(Epigaea repens),* Goldthread *(Coptis groenlandica),* and Pipsissewa *(Chimaphila umbellata),* together with a good many ferns and fern allies. There is a mature forest of Eastern Hemlock *(Tsuga canadensis)* and another of White Pine *(Pinus strobus)*. For good measure, add a salt marsh and a small bog. And of course both sides have the inevitable rocky shorelines.

Ferry Beach State Park

If you drive back to Portland and continue south along the coast for about the same distance, you begin to make the transition from rocky to sandy seacoast. In fact, the venerable resort of Old Orchard Beach owes its name and much of its popularity to a 7-mile long strand of gleaming white sand.

South of the town on S.R. 9 is the 100-acre Ferry Beach State Park, where visitors taking the walkway down to the shore of Saco Bay can enjoy dune plants like Sweetfern *(Comptonia peregrina)*, Beach Pea *(Lathyrus japonicus)*, and Woolly Beach Heather *(Hudsonia tomentosa)*. Also prevalent is Northern Bayberry *(Myrica pensylvanica)*, which has the capability of enriching the soil by fixing nitrogen in its root nodules—a property more generally associated with leguminous plants. Among those that are helping to stabilize the dunes are such disparate species as Beach Grass *(Ammophila breviligulata)*, Poison Ivy *(Rhus radicans)*, and Pitch Pine.

In addition, the park encompasses deciduous woodlands, a pond, and a swamp. As might be expected, the latter has Highbush Blueberry *(Vaccinium*

NORTHERN BAYBERRY
Myrica pensylvanica

corymbosum), Maleberry *(Lyonia ligustrina),* and Winterberry Holly *(Ilex verticillata),* but it is most notable for a sizable stand of Black Gum *(Nyssa sylvatica),* a tree of much more southern distribution. All of these components can be explored by means of a series of interlocking foot trails.

"THE CAPE"

If glaciation played an important role in the shaping of the rugged northeast coast, it contributed even more directly to the formation of the sandy shores between southern Maine and Long Island. The materials that made up these shores began as rocky debris scoured from the New England mountains and left behind by the melting ice mass as moraines and include finer sediments carried still farther out by glacial rivers. It then remained for the action of waves, currents, and winds to slowly mold these deposits into the land forms that are so familiar to us now—including the long, gracefully curved outer prong of Cape Cod that reminds some of a beckoning forearm, others of a scorpion's tail.

At its narrowest point this spit is only a mile across, and nowhere does it exceed 4.5 miles in breadth, yet its physiography is remarkably varied. The east-facing oceanfront has incredibly long stretches of beach, shifting dunes, and steep bluffs. These contrast with the opposite side, where marshes and tidal flats rim the more placid Cape Cod Bay. The interior has numerous kettle ponds, swamps and bogs, heaths, and wind-whipped forests of oak and pine.

Whereas Maine has a predominantly rockbound coast interrupted here and there by sandy beaches, the northern shore of Cape Cod is just the reverse. The contrast is dramatically evident in the complete absence of visible plant life from the intertidal zone of the beaches, where the unstable sands do not afford a base firm enough for even the most tenacious seaweeds to attach their holdfasts.

This difficulty in becoming established is shared by some of the higher plant forms that attempt to grow at considerable distances back from the upper tidal limits. There they are confronted by strong salt-laden winds, temperature extremes, and a supply of fresh water that is scant because of its rapid percolation through the porous sand. A few grasses are able to survive these harsh conditions, and since they not only help to hold the sand in place but actually contribute toward the formation of dunes by acting as obstacles to windblown drifts, they are planted exclusively as dune stabilizers. It should be noted here that although these specialized grasses are capable of withstanding some of the most furious onslaughts of nature, they may quickly succumb to seemingly minor human disturbances, and the loss of their protection leaves the dunes themselves vulnerable to destructive erosion. It should therefore be one's policy never to walk or drive on a sand dune if it can be avoided, but to use elevated boardwalks, stairs, and similar structures where they have been provided.

The most successful sand binder is Beach Grass, which persistently produces new plants despite inundation by blowing sand. Under less violent conditions near the crests of the ridges, Beach Grass is joined by Beach Pea, a very different kind of plant but one that also helps to stabilize the dunes. It does so by sending out runners over the surface and coiling its tendrils around tufts of grass. In common with many other leguminous plants, Beach Pea has roots that form nitrogen-fixing nodules, thereby contributing to the creation of fertile soil from the sterile sands.

Another common plant in this zone is *Artemisia stelleriana,* a composite introduced from Asia and well-known to gardeners as Dusty Miller. The velvety covering of silver hairs that gave it that name serves it well in this environment by holding down the rate of evaporation from its stems and leaves—the same adaptation that permits its sister plants, the Sagebrushes, to survive the dryness of our western deserts and plains.

Where two rows of dunes lie parallel, the protected secondary ones—as well as the backsides of the primary dunes—offer a haven to some only slightly less rugged plants, including the coarse Seaside Goldenrod; Salt Spray Rose, another native of the Orient with splendid large white to deep pink flowers; and yellow-spangled mats of Woolly Beach Heather. Notable among the woody plants are Northern Bayberry and Beach Plum *(Prunus maritima),* the sources of candles and jellies that many visitors take home from the New England seacoast.

The vegetation to be found in the low areas, or swales, between parallel dunes will vary with moisture conditions. Drier hollows will foster thickets of Highbush Blueberry *(Vaccinium corymbosum),* but if brackish water is present Groundsel Tree *(Baccharis halimifolia)* will flourish. Substantial quantities of water often accumulate in these depressions to create acidic Sphagnum bogs—ideal habitat for Cranberries *(Vaccinium macrocarpon)* as well as various orchids and insectivorous plants.

Moving inland from the secondary dunes toward the uplands, you leave behind the highly adapted, exclusively maritime species and begin to see more and more of those associated with the interior: Pitch Pine, Eastern Red Cedar *(Juniperus virginiana),* Wild Cherry *(Prunus serotina),* Red Maple *(Acer rubrum),* Black Gum, and others.

Where accumulations of sediment eroded from soft bedrock are protected from wave action, salt marshes develop. Although spared the violence of the open sea, they are subject to wide fluctuations in temperature and in the volume and salinity of the water that washes over them. Salt marsh vegetation consists mainly of grasses, often with a definite zonation of species indicative of their peculiar requirements. Important components are Cordgrass *(Spartina alterniflora)* and Salt-meadow Hay *(S. patens),* which early settlers utilized for roof thatch and cattle fodder, respectively. Somewhat less tolerant of inundation is *Juncus gerardi,* known as Black Rush because of the contrast between its dark leaves and the brighter greens of adjacent grasses. Projecting from the surface of the high marsh are the translucent, turgid stems of Glasswort (*Salicornia* spp.). Some flowering plants venture down to the edges of the marshes but keep clear

CRANBERRY
Vaccinium macrocarpon

of salt water flooding; among these are Sea Lavender *(Limonium nashii),* a diffusely branched plant with tiny flowers, and at the other extreme Swamp Rose Mallow *(Hibiscus palustris),* which boasts one of the biggest blossoms of any of our native wildflowers.

Cape Cod National Seashore

Of great interest to tourist and naturalist alike is the Cape Cod National Seashore, which spans the coast from Chatham, site of an early lifesaving station, past the cliff top from which Marconi sent the first transatlantic radio message, to Provincetown and the spring where the Pilgrims paused to take fresh water aboard the *Mayflower* before putting in at Plymouth.

There are two visitor centers—Salt Pond at Eastham, in the crook of the "elbow," and Province Lands, at the northern extremity near Provincetown. Between these two points, U.S. 6 in conjunction with numerous side roads provides access to self-guiding nature trails, bicycle trails, bridle paths, and locations for fishing, swimming, and surfing.

The mile-long Nauset Marsh Trail begins alongside Salt Pond, which was originally a freshwater kettle formed when a huge block of ice left behind by the retreating glaciers melted, creating a depression, but is now salty from the intrusion of seawater. It then skirts a portion of the extensive salt marsh before turning inland. Here areas formerly used for farming and recreation are being reclaimed by native pioneer plant species, led by Eastern Red Cedar *(Juniperus virginiana).* Nearby are the Buttonbush Trail, with features to aid the vision-impaired, and a 1.6-mile bicycle path leading to the ocean.

South of Salt Pond is the Fort Hill Trail, where evidence can be seen of the human occupation of this fertile area, which began with the Nauset Indians and ended only in the 1940s. Part of the trail leads through a Red Maple swamp *(Acer rubrum),* which includes a stand of Black Gum, or Tupelo *(Nyssa sylvatica).*

The Atlantic White Cedar Swamp Trail is reached from the parking area at the Marconi Station Site (the entrance road is on the east side of U.S. 6 approximately 5 miles above Salt Pond). The tree for which it is named, *Chamaecyparis thyoides,* is largely a southern coastal species, and its occurrence here makes this swamp a botanically significant feature of the Cape.

Behind the dunes, the trail heads away from the ocean through a scrubby growth of Bear Oaks *(Quercus ilicifolia)* and Pitch Pines, trees pruned by the savage winds to pygmy size but found to be 75 years old and more. Huddling still closer to the sand are tufts of Broom Crowberry *(Corema conradii)* and Golden Heather *(Hudsonia ericoides).* As the land slopes gradually downward, there are signs of increased moisture and richer soils. White and Black Oaks (*Q. alba* and *Q. velutina*) over Wild Sarsaparilla *(Aralia nudicaulis),* Canada Mayflower *(Maianthemum canadense),* and Wintergreen. Then these give way to Red Maples and such shrubs as Sweet Pepperbush *(Clethra alnifolia),* Inkberry *(Ilex glabra),* and Swamp Azalea *(Rhododendron viscosum).* Eventually you come to the swamp itself and the circular boardwalk that takes you through the dense stand of Atlantic White Cedars—stately relics of warmer times.

Cape Cod is probably more closely identified with the large-fruited Cranberry *(Vaccinium macrocarpon)* than with any other plant. The reason for this successful association becomes clear when you see wild vines growing under conditions that respond perfectly to their unusual and complex requirements—in low, boggy swales where they receive abundant fresh water and are shielded by dunes that protect them from the force of strong winds but permit gentler breezes to wash them over with fine sand from time to time. This can be observed near the Province Lands Visitor Center on Race Point Road in Provincetown, but it may be even more instructive to visit a cultivated bog like the one on North Pamet Road east of Truro. This centuryold facility, which occupies the site of an old maple swamp, has been preserved by the National Park Service, which has also built a short trail into the bog; because of frequent flooding, you should wear waterproof footgear. The cranberries are in flower during June and July, and bear fruit from September to October.

At Pilgrim Heights in North Truro there are two short (three-quarter-mile) circular interpretive trails. The Small's Swamp Trail ends at a sheltered glacial kettle named for a Thomas Small, who set up farming here in the 1860s. Passing through a Pitch Pine–Black Oak forest, it emerges onto a sandy ridge overlooking a freshwater marsh with the Atlantic Ocean in the distance. The dunes are carpeted with Bearberry *(Arctostaphylos uva-ursi),* whose fruits are vivid red but are dry and mealy inside, and are called hog cranberries locally. The trail then turns to follow the edge of the swamp. Starting from the same point is the Pilgrim Spring Trail, which leads to the place where it is believed a small band of Pilgrims from the *Mayflower* had their first taste of water in the New World.

The south-to-north series of nature trails culminates in the Provincelands Beech Forest. The handsome American Beech *(Fagus grandifolia)* apparently found the soil and climate of the Cape well suited to its needs, although wind erosion resulting from various human activities has taken a heavy toll. The trail consists of two loops joined end to end. The first goes through mixed hardwoods in association with Inkberry, Beach Plum *(Prunus maritima),* and Sheep Laurel *(Kalmia angustifolia),* and encircles shallow dune ponds embellished with Water Lilies. The other loop swings through the Beech woods, where attractive wildflower species include Starflower *(Trientalis borealis)* and Spotted Wintergreen *(Chimaphila maculata).*

PLUM ISLAND

One winter in the 1970s, people by the thousands came to peer at a square foot of snow near Newburyport in the northeastern corner of Massachusetts. The focus of their attention was a rare bird that very few have ever seen below the Arctic Circle—a single, solitary, individual Ross's gull resting after apparently having strayed far off course. Although an aberrant occurrence not likely to be repeated, the incident served as a dramatic reminder that neighboring Plum Island has long been famous for the variety and numbers of its bird life, and inevitably renewed interest in the physiography and vegetation of the region.

Plum Island is the largest barrier beach north of Cape Cod, and one of the best preserved. The basis of its formation was a group of drumlins, peculiar whale-shaped hills that were formed of glacial debris and after partial submergence by the sea, appeared as separate islands offshore. Ocean and river currents then deposited sand around them until eventually they were connected to each other to form the foundation of a barrier beach.

Most of the island, as well as acreage on the mainland, is contained within the Parker River National Wildlife Refuge. It is easily reached by exiting from I-95 to S.R. 113 east, then following S.R. 1A and signs for the Refuge Headquarters.

A popular feature is the Hellcat Swamp Nature Trail. Starting from parking lot No. 4, this 2-mile interpretive trail introduces the visitor to the island's principal plant communities. The eastern portion of the trail skirts two freshwater swamps with Speckled Alder *(Alnus rugosa),* Serviceberry *(Amelanchier canadensis),* Willows (*Salix* spp.), and Red Maples growing over a thick understory of Winterberry, Maleberry, Highbush Blueberry, Arrowwood *(Viburnum recognitum),* Steeplebush *(Spiraea tomentosa),* Virginia Creeper *(Parthenocissus quinquefolia),* and many others.

On the back dunes there are Northern Bayberry, Beach Plum, and Black Cherry, with Blackberries (*Rubus* spp.), Wild Roses (*Rosa* spp.), and the Eurasian Morrow's Honeysuckle *(Lonicera morrowi)* forming thickets. Another exotic, Austrian Pine *(Pinus nigra),* has been planted as a dune stabilizer. In quiet areas back from the ocean there are patches of False Heather *(Hudsonia tomentosa)* and interdunal swales with mats of Cranberry vines.

The western part of the trail leads to a wildlife observation blind, then returns over a loop that passes by a freshwater marsh dominated by Cattails (*Typha* sp.), Reed *(Phragmites australis),* and Purple Loosestrife *(Lythrum salicaria).*

3

Botanizing for New Yorkers

ITS INITIAL attraction was a magnificent natural harbor, and this discovery was dutifully reported by Giovanni da Verrazano, Estevan Gomez, Henry Hudson, and other early explorers to their respective patrons. When the Dutch established a trading post in the 17th century, their main concern was bartering with the Indians for furs. Later, as an actual settlement began to take shape at New Amsterdam, the colonists became aware of the abundance of fish and game and wild fruits, but to their eyes the land was a resource to be tamed and soon they were bringing domestic animals, poultry, and familiar plants over from the homeland. Gradually their little farms spread inland, but even after the surrender to England the countryside beyond the village that was then known as New York remained pretty much in its original state.

It was not until the mid-1700s that European scientists began to evince serious interest in the flora of the region. If, in looking at New York City today, we find it hard to envision botanists like Peter Kalm exclaiming over the wealth of strange new plant species and gathering up huge collections of specimens for his mentor, the great Linnaeus, we may be forgiven some of our skepticism, for the intervening years have seen the landscape steadily obliterated until now the city seems devoid of any trace of its natural heritage.

This is not true, of course, for there still is lots of green space within the boundaries of the five boroughs. The Ramble, a singularly resistant piece of woodland, survives in Frederick Law Olmsted's Central Park, which is itself Manhattan's centerpiece. Van Cortlandt Park, at the upper end of the Bronx, is one of the city's largest open areas. Shared by Brooklyn and Queens, and next door to one of the world's busiest airports, are 12,000 acres of wetlands constituting the Jamaica Bay Wildlife Refuge. And Staten Island—the "forgotten borough" until the opening of the Verrazano Narrows Bridge—has the William T. Davis Wildlife Refuge.

For thousands of New Yorkers, these and similar facilities conveniently located within the city limits provide a welcome refuge from the hectic world of commerce. To those with a more than passive interest in nature, however, they are often found to be inadequate—as, for example, where native vegetation has become so corrupted by the encroachment of foreign species that it can no longer serve as a model for study. The answer is to find other sites in the suburbs and beyond; there they are more likely to be free from the stifling influences of an urban environment. This is one of several reasons for the rapidly growing tendency of city dwellers to make frequent but short recreational trips, often within a 50-mile radius of their homes. Fortunately, the possibilities do not diminish as you move farther out from the city; in fact, the choices become even more varied. There literally are dozens of natural areas that can be enjoyed comfortably on a day's outing, and what follows is a mere sampling taken more or less at random.

WARD POUND RIDGE RESERVATION

The pride of Westchester County's Department of Parks is the big, sprawling Ward Pound Ridge Reservation. From an assemblage of fields and woodlots belonging to more than 30 farms, it has been expanded to 4,700 acres of beautiful rolling landscape ideally suited to nature study and outdoor recreation.

It is easily reached from I-684 by exiting at Katonah onto S.R. 35 east for about 4 miles, then turning south for 0.1 mile to the entrance road on the left.

Ward Pound Ridge has a bewildering complex of trails aggregating more than 35 miles, and a good idea is to continue on for a short distance beyond the entrance station to the Trailside Nature Museum. Here you can inspect a detailed map, obtain information about trail conditions and plants that may be of particular interest, and choose one or more walks. You can hardly go wrong: Most trails are through wooded areas and are at their best—although they may be wet—in early spring, but there are plenty of sunlit places in which to enjoy summer and fall flowers as well.

You might decide to drive through the park on Boutonville Road to Kimberley Bridge, where there are Bottle Gentians *(Gentiana andrewsii)* along with a wealth of Asters in the fall, and such plants as Knapweed *(Centaurea dubia)* and Field Milkwort *(Polygala sanguinea)* somewhat earlier. The River Trail, which leaves northward from here, is excellent for spring wildflowers, some of which are Spring Beauty *(Claytonia virginica)*, Wild Columbine *(Aquilegia canadensis),* Trout Lily (*Erythronium* americanum), Bloodroot *(Sanguinaria canadensis),* Dwarf Ginseng *(Panax trifolium),* Showy Orchis *(Galearis spectabilis),* and Round-lobed Hepatica *(Hepatica americana).* An interesting assortment of blue Violet species has been noted along this trail: Downy, Early Blue, Dog, Long-spurred, and Marsh (*Viola fimbriatula, V. palmata, V. conspersa, V. rostrata,* and *V. cucullata*), as well as Common Blue *(V. papilionacea).*

Another section worth exploring is south of the parking lot at the end of Michigan Road (which branches off near the entrance booth). Although wet ground may present some problems early in the year, it provides such interesting plants as Skunk Cabbage *(Symplocarpus foetidus)* and Water Plantain *(Alisma triviale),* and stunning displays of Marsh Marigolds *(Caltha palustris),* to say nothing of ferns, of which 24 species have been documented within the boundaries of the reservation. Also found in moist situations are such trees as Pussy Willow *(Salix discolor),* Witch Hazel *(Hamamelis virginiana),* Ironwood *(Carpinus caroliniana),* and Spicebush *(Lindera benzoin);* and herbaceous flowering plants ranging from Canada Lily *(Lilium canadense)*—including the red form—White Turtlehead *(Chelone glabra),* and Swamp Candles *(Lysimachia terrestris)* to the modest Grove Sandwort *(Arenaria lateriflora)* and the inconspicuous Golden Saxifrage *(Chrysosplenium americanum).*

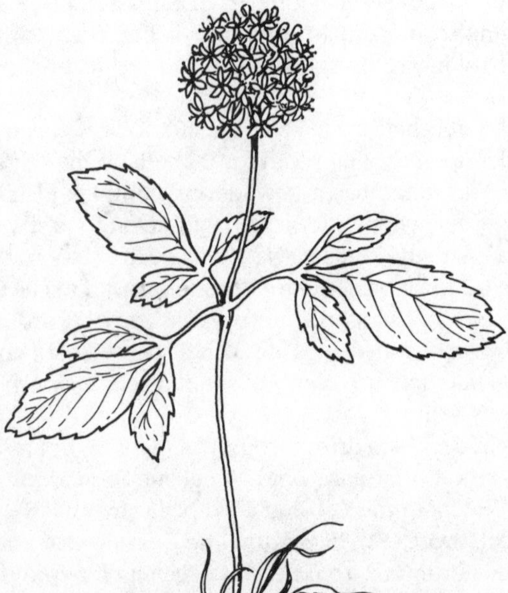

DWARF GINSENG
Panax trifolium

Symplocarpus foetidus

WESTMORELAND SANCTUARY

Although New York's Westchester County is one of the nation's most affluent, private interests have set aside a surprisingly large number of tracts of valuable land as nature preserves and have opened them up to the public for study and enjoyment. One such entity is Westmoreland Sanctuary, a 625-acre enclave with 15 miles of trails and an extensive program of educational activities.

The entrance is on the east side of Chestnut Ridge Road, which parallels I-64 and is accessible either from S.R. 172 east of Mount Kisco or from S.R. 22 north of Armonk.

A frame church, originally built at Bedford Village in 1782, was dismantled and rebuilt here, and serves as a nature center and museum. Nearby are some large Sugar Maples *(Acer saccharum),* where a demonstration of sugaring off takes place each spring; an old cemetery; and the starting point for walking through the preserve.

The trail goes gradually downhill through conifers, of which White Pine *(Pinus strobus)* and Eastern Hemlock *(Tsuga canadensis)* are the most abundant here. Along the edges of this path, as well as others, evergreen Grape Ferns *(Botrychium dissectum)* are frequently seen. At Bechtel Lake the trail goes right, paralleling the lake. Sharing the muddy shores with sedges is Low Cudweed *(Gnaphalium uliginosum),* and at the southern end a collection of native ferns was established a number of years ago.

From this point you can take the Chickadee Trail, which passes by some imposing rock ledges, and, by bearing right, pick up Lost Pond Trail. This circles Lost Pond, which is a tranquil spot perfectly suited to having a trail lunch or

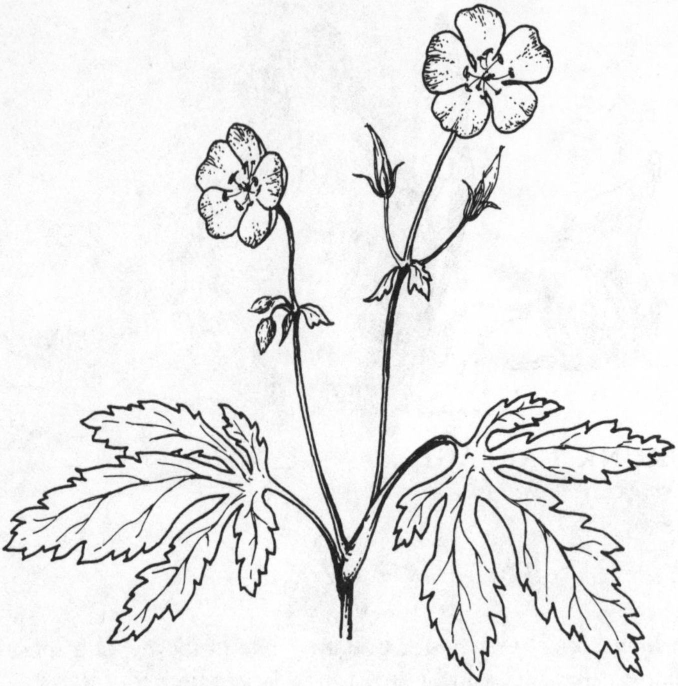

WILD GERANIUM
Geranium maculatum

simply resting. The shortest route back would be to retrace your steps, but there are several alternatives to choose from by consulting the trail map.

It has been a number of years since I walked the trails at Westmoreland, and while many of the wildflowers and other plants are still vivid in my memory, it is difficult to recall their exact locations. I know, for example, that there were Pink Moccasin Flowers *(Cypripedium acaule),* the waxy white flowers of Shinleaf *(Pyrola elliptica),* and spikes of Rattlesnake Plantain *(Goodyera pubescens)* arising from neatly veined rosettes, and I suspect that they had all come up through the brown needle cover under pines. Red-and-yellow Columbine *(Aquilegia canadensis)* and Solomon's-seal *(Polygonatum pubescens)* probably had the big rocks for a background, while the little Pennsylvania Bitter Cress *(Cardamine pensylvanica)* must have been at one of the ponds or brooks. Others, like Wild Geranium *(Geranium maculatum),* Wood Betony *(Pedicularis canadensis),* two of the Bellworts (*Uvularia sessilifolia* and *U. perfoliata*), White Baneberry *(Actaea pachypoda),* and the handsome Pinxter Flower (then known as *Rhododendron nudiflorum* but since renamed *R. periclymenoides),* could have been almost anywhere in the woods.

September was interesting for fungi, especially on the loop trail beside Bechtel Lake. There were odd-looking mushrooms like the Pine Cone Fungus *(Strobilomyces strobilaceous),* but most impressive were the ones with jewellike colors: the rose red *Russula emetica,* the lilac *Cortinarius alboviolaceus,* the tiny trumpets of Vermilion Chanterelle *(Cantharellus cinnabarinus),* and the yellow-orange brackets of Sulphur Polypore *(Polyporus sulphureus).*

AUDUBON CENTER IN GREENWICH

To suggest that New Yorkers go to Connecticut for botanizing might seem irrational, until you remember that the southwestern corner of the state comes within a mere 7 miles of the Hudson River shoreline. This puts the Audubon Center in Greenwich less than 30 miles from the city, and closer than some sites that are located in its own outer suburbs.

The center is actually in the countryside 8 miles above Greenwich. An easy way to get there is to take New York Highway 22 east from I-64 (at Armonk) and turn right at the first traffic light (this is marked as Route 433 but changes at the state line to Connecticut Highway 128, or Riversville Road). In about 2 miles, at John Street, make an acute left-hand turn through the gate.

The Audubon Center occupies 280 acres of diverse habitat, bisected from north to south by the Byram River. Evidences of earlier settlement include the stone fences of early 18th century farms and a 100-year-old millpond.

A 1.5-mile loop trail begins just below the interpretive building. By following the signs for the Byram River and Mead Lake you will come to a small pond with a stone spring house. From here the Discovery Trail joins Riverbottom Road as far as the dam, which was built to run a waterwheel for powering a sawmill. At this point the circular Lake Trail takes over.

As the path traces the eastern shore of the lake where blinds for observing bird life have been built, small moisture-loving plants become evident. Close to the trail here and elsewhere on the perimeter you will see the geraniumlike leaves of Water Pennywort *(Hydrocotyle americana),* Bugleweed *(Lycopus virginicus),* dainty Marsh Bellflowers *(Campanula aparinoides),* and the prostrate Mermaid Weed *(Proserpinaca palustris).* All have such unassuming flowers that those of creeping Moneywort *(Lysimachia nummularia)* and even Mad-dog Skullcap *(Scutellaria lateriflora)* seem ostentatious by comparison. Other plants of more ample proportions can be seen in waters of the lake or around its edges: Water Lilies *(Nymphaea odorata),* Blue Flags *(Iris versicolor),* and Cattails (*Typha* sp.), for example. These will be more apparent as the trail turns to cross the inlet above the lake, then cuts across an open marsh and down the opposite shore. Woody plants to look for here are Sweet Pepperbush *(Clethra alnifolia),* Poison Sumac *(Rhus vernix),* and Winterberry *(Ilex verticillata).* This side is especially rich in spring wildflowers, too.

Near the foot of the lake there is a short spur off to the left where a small colony of Round-leaved Sundew *(Drosera rotundifolia)* is nestled in a bed of Sphagnum moss. Although this is considered unusual here, another insectivorous plant—Common Bladderwort *(Utricularia vulgaris)*—is relatively plentiful at Mead Lake. Shortly you will meet the other leg of the Discovery Trail, which leads uphill past an old apple orchard to the starting point.

Two other sections of the sanctuary can be cited for their vegetation. One is an old meadow that lies east of Riversville Road near the entrance, now abandoned to Little Bluestem *(Schizachyrium scoparium),* Northern Bayberry *(Myrica pensylvanica),* Gray Birch *(Betula populifolia),* and the like, while the bordering woods have had to deal with a rampant invasion of Oriental Bittersweet *(Celastrus orbiculatus).* As must be expected, the complement of wildflowers is subject to rapid change, but at times such species as Purple Milkweed *(Asclepias purpurascens),* Indian Hemp *(Apocynum cannabinum),* and Blazing Star *(Liatris spicata)* have been displayed.

The other section is a swamp at the northern end, composed principally of Red Maple *(Acer rubrum)* in association with Black Ash *(Fraxinus nigra).* Of special interest here is the occurrence of Boott's Shield Fern *(Dryopteris boottii)* with the more common Crested Shield Fern *(D. cristata).*

Under the same administration as the Audubon Center is the nearby Audubon Fairchild Garden, a beautiful 127-acre tract, mostly forested. It is home to an even greater number of wildflowers, but this is due in part to the fact that many pharmaceutical plants (reflecting the business interests of the founder, Benjamin Fairchild) have been introduced, and for this reason it is listed in the

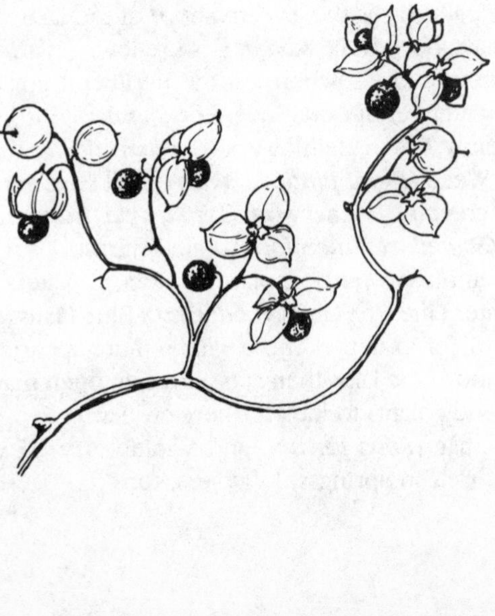

BITTERSWEET
Celastrus orbiculatus

Appendix with other botanic gardens and arboretums. To reach it from the Audubon Center, simply continue south on Riversville Road and turn left on North Porchuck Road for one-half mile; the entrance gate is on the right.

THE PALISADES

One of the country's most distinctive geological formations stands on the west bank of the Hudson River opposite New York City. Appropriately named "the Palisades," it is a continuous wall of rust brown rock rising vertically from the shoreline to over 500 feet, and is most striking from the George Washington Bridge northward for about 25 miles.

These spectacular cliffs represent the edge of a molten rock mass that cooled underground, later to be exposed by erosion. Today the forces of nature still continue their attack on these ramparts, causing huge blocks and splinters to break off and crash down, adding to the already enormous talus slopes at the base.

In the last century there were those who saw the Palisades as a resource to be quarried, and the prospect of eventual defacement brought outraged protests from others who were determined to preserve their scenic grandeur. Legislative action, combined with the generosity of wealthy families like the Rockefellers, Morgans, and Harrimans, culminated in the acquisition of land in New York as well as New Jersey, which has since grown to a total of 80,000 acres and is administered jointly by the two states through the Palisades Interstate Park Commission. Included under its jurisdiction are not only the Palisades themselves but Bear Mountain, Harriman, and a number of other smaller parks.

In 1958 a scenic drive known as the Palisades Interstate Parkway was completed along the cliff top from the George Washington Bridge to the New York state line, then curving across the Hudson highlands to terminate at Bear Mountain. It is especially beautiful in early May when the Dogwoods are in bloom, but whatever the season one should not fail to turn off at Rockefeller Lookout near Englewood Cliffs and walk to the brink for a breathtaking view of the bridge with the distant Manhattan skyline forming a backdrop.

Between the parkway and the Hudson River, the New Jersey Palisades are served by a network of foot trails, some along the top of the cliffs, some tracing the shoreline, and still others connecting the two by stairs. One place to try them out is at exit 2 (near Alpine), where a road leads down past the Park Commission headquarters (which is *not* a visitor center). Turn left at the fork near the bottom, and park under a row of old Sycamores at the boat basin.

Walking north you will pass a white building, which was formerly a tavern and served as Lord Cornwallis's headquarters in November 1776. To the left of the picnic shelter a road climbs gradually uphill, then levels off; this is the beginning of the Old Alpine Trail, which the British used in a desperate but unsuccessful attempt to cut off General Washington's march to Trenton. By staying on this road (and ignoring the stairs on the left, which lead back to the top), you will be able to walk the shore trail all the way to the state line—or just as far as you wish.

These woods are predominantly Sugar Maple, although there certainly are enough Oaks, Hickories, Tulip Trees, Beeches, and Sassafras to achieve variety. Farther on, in a moister environment, Eastern Hemlocks take over, and the understory is heavy with Maple-leaved Viburnum *(Viburnum acerifolium).* Wildflowers are those that accommodate well to rocky situations, such as Wild Columbine, Bloodroot, Jack-in-the-Pulpit *(Arisaema triphyllum),* and Rue Anemone *(Anemonella thalictroides);* later these are supplanted by Horse Balm *(Collinsonia canadensis),* Panicled Hawkweed *(Hieracium paniculatum),* Spanish Needles *(Bidens bipinnata),* and Indian Tobacco *(Lobelia inflata).* Herb Robert *(Geranium robertianum)* grows out of crevices in the rock walls.

A short course of stone steps on the right leads down to the river. Just beyond a patch of Bladdernut bushes *(Staphylea trifolia),* note the old shells protruding from the bank, indicating that these stairs were cut through an old Indian midden dating back to a time when the salinity of the Hudson at this point was high enough for oysters. Down at the shore, Fox Grape vines *(Vitis labrusca)* are luxuriantly festooned over the Basswoods and Witch Hazels. And there are lavender-flowering Princess Trees *(Paulownia tomentosa),* the progeny of specimens imported from China to embellish the grounds of fine old estates that once lined the river.

ROCKLAND LAKE

A well-developed unit of the Palisades Interstate Parks system, Rockland Lake State Park is located in New York between U.S. 9W and the Hudson River, just above the western end of the Tappan Zee Bridge near the town of Congers.

A popular feature is its nature center, which includes live exhibits of native fauna and nature trails with interpretive signs. To reach this facility, go in at the *north* entrance and park in parking field no. 6, which faces the glacially formed lake. Walk to the left on the path past the boat rental dock; this curves to the right and leads very shortly to the museum building and trailhead.

The Woodland Swamp Trail begins at a wooden bridge and enters a Red Maple–White Ash *(Acer rubrum–Fraxinus americana)* swamp with Spicebush, Sweet Pepperbush, and Royal Fern. After passing briefly through a section of dry hardwood forest, the path crosses a small stream lined with the arching stems of Swamp Loosestrife *(Decodon verticillatus).* This feeds into the lake, where glimpses can be had of large beds of Arrow Arum and tall spikes of Purple Loosestrife, some entwined by the orange stems of Dodder vines (*Cuscuta* sp.). Returning to the wooded swamp, the trail skirts Red Osier Dogwood *(Cornus stolonifera)* and Maleberry *(Lyonia ligustrina)* over patches of Sensitive Fern.

The second trail is labeled "Lakeside Bog," but this is now pretty well filled in with shrubby growth, including Speckled Alder *(Alnus rugosa),* Arrowwood *(Viburnum recognitum),* Winterberry Holly, Swamp Azalea *(Rhododendron viscosum),* and Highbush Blueberry. Of particular interest, especially to the many visitors who have been wondering what it looks like, are several labeled specimens of Poison Sumac *(Rhus vernix)*—all near enough to be seen clearly but growing well beyond reach.

The vegetation along the paved path leading back to the parking lot is as interesting as that along the trails. For example, the presence of Box Elder, or Ash-leaved Maple *(Acer negundo),* in proximity to White Ash affords an opportunity to compare their foliage. Three unoffending species of Sumac are found here: Smooth, Staghorn, and Winged (*Rhus glabra, R. typhina,* and *R. copallina*). The metallic blue fruits of Silky Dogwood *(Cornus amomum)* furnish a contrast with the white berries of Red Osier *(C. stolonifera).* Climbing plants include the pernicious Oriental Bittersweet as well as Hedge Bindweed *(Convulvulus sepium),* Climbing Hempweed *(Mikania scandens),* and Groundnut *(Apios americana).* And summer brings vividly colored wildflowers like Swamp Milkweed *(Asclepias incarnata),* Jewel Weed *(Impatiens capensis),* and Great Blue Lobelia *Lobelia siphilitica).*

GREAT SWAMP

Almost within sight of the New York City skyline is a 7,850-acre natural area where wood ducks breed by the thousands; where deer, red fox, raccoon, and muskrat are abundant; and where a variety of habitats combine to foster a diverse flora. This unique haven in heavily urbanized north-central New Jersey, due west of midtown Manhattan, is known as the Great Swamp. A large portion was designated a National Wildlife Refuge in 1960 after a groundswell of public opposition to a proposed jetport resulted in the raising of a million dollars with which the federal government purchased the initial tract.

The Great Swamp itself began as part of prehistoric Lake Passaic, which was formed out of the meltwater of the retreating Wisconsin glacier. Eventually most of the water drained away, and the resulting basin is now a conglomeration of woods, grasslands, swamps, and ponds.

The refuge headquarters is reached by taking exit 26 (near Basking Ridge) from I-287 onto North Maple Avenue, then turning left on Madisonville Road and right on Pleasant Plains Road. The focal point for many visitors is the wildlife observation center, east of the headquarters off Long Hill Road, where large blinds have been provided for watching and photographing wildlife.

A short walk through fragrant bowers of Sweet Pepperbush brings you to a boardwalk leading to the smaller bird-watching blind. The ponds are bordered in both Common and Narrow-leaved Cattails (*Typha latifolia* and *T. angustifolia*), and feature Arrow Arum *(Peltandra virginica),* Pickerel Weed *(Pon-*

tederia cordata), and Water Lilies. Shoreside shrubs include Buttonbush *(Cephalanthus occidentalis)* and Steeplebush *(Spiraea tomentosa).* After passing through a wooded section, the path ends before a profuse expanse of Water Pepper *(Polygonum hydropiperoides)* and Tearthumb *(P. arifolium* and *P. sagittatum).*

The boardwalk to the larger blind transits a pleasant, shady Red Maple swamp with a luxuriant growth of ferns, of which Cinnamon *(Osmunda cinnamomea),* Royal *(O. regalis),* and New York *(Thelypteris noveboracensis)* are the most abundant. Sweet Gum trees *(Liquidambar styraciflua)* with their attractive star-shaped leaves are plentiful here although approaching the northern limit of their range. Clumps of Water Plantain *(Alisma subcordatum)* sprinkled with tiny white flowers arise out of dark pools, while Canada Mayflowers *(Maianthemum canadense)* carpet the higher and drier ground. Out in the sun near the blind there are banks of orange Jewel Weed overlooking Pickerel Weed and Arrowhead *(Sagittaria latifolia)* in the open water.

In 1968 the eastern part of the refuge was set aside as a Wilderness Area, but foot travel is in conformity with that designation and there are ample opportunities for exploring along 8 miles of trails. One access point is south of the wildlife observation center on White Bridge Road, from which an old road proceeds in a generally northward direction. Now abandoned to nature, this wide swath is becoming a meadow, especially colorful in summer and fall with a mixture of native and alien wildflowers: Queen Anne's Lace *(Daucus carota).* Thin-leaved Coneflower *(Rudbeckia triloba),* Deptford Pink *(Dianthus armeria),* Sweet Joe-Pye-Weed *(Eupatorium purpureum),* Boneset *(E. perfoliatum),* Birdfoot Trefoil *(Lotus corniculatus),* and many others.

More remarkable is the great variety of trees and shrubs that make up the woods on either side. They include Sour Gum *(Nyssa sylvatica),* Black Walnut *(Juglans nigra),* Sassafras *(Sassafras albidum),* Flowering Dogwood *(Cornus florida),* Quaking Aspen *(Populus tremuloides),* Spicebush, Smooth and Winged Sumac, and Arrowwood, to name a few. Openings occur at intermittent swampy patches where such plants as Swamp Azalea, Maleberry, and Sensitive Fern *(Onoclea sensibilis)* occur.

Immediately west of the refuge, the Somerset County Park Commission maintains an environmental education center. This can be reached by taking the same exit from I-287 but continuing on North and South Maple Avenues and turning left on Lord Stirling Road.

There is an 8.5-mile network of trails here, starting out between two ponds rimmed with Purple Loosestrife *(Lythrum salicaria),* a brilliant European perennial that is becoming increasingly abundant in the wetlands of the northeastern United States, together with Blue Vervain *(Verbena hastata),* Silky Dogwood, and Groundnut. Aromatic mints are common: Wild Mint *(Mentha arvensis),* Peppermint *(M. piperata),* and Narrow-leaved Mountain Mint *(Pycnanthemum tenuifolium).*

Entering an oak-hickory-maple wood, the path crosses a small brook lined with Gray Birches and brightened in late summer by a stand of Cardinal Flower *(Lobelia cardinalis).* Farther along there are fine specimens of American Beech *(Fagus grandifolia)* and Shagbark Hickory *(Carya ovata),* and these are followed by a boardwalk between masses of Netted Chain Fern *(Woodwardia areolata)* on one side and Lady Fern *(Athyrium filix-femina)* on the other.

A treeless opening is dominated by Goldenrods (*Solidago* spp.), Ironweed *(Vernonia noveboracensis),* and Steeplebush, with Purple Gerardia *(Gerardia purpurea)* and Field Milkwort among the smaller trailside plants.

Adjoining the extreme eastern edge of the refuge proper is the Great Swamp Outdoor Education Center operated by the Morris County Park Commission. The entrance is on the west side of Southern Boulevard in Chatham.

Although this facility has only a modest 1-mile interpretive trail, it traverses richly varied terrain and is especially interesting in the spring. It begins with a dry deciduous forest on one side, where Pink Moccasin Flower, Starflower *(Trientalis borealis),* Wild Sarsaparilla *(Aralia nudicaulis),* and Spring Beauty *(Claytonia virginica)* are among the wildflowers, and a Red Maple swamp on the other. Farther along, the prevalence of Gray Birches signifies the early stages of the transition from an old field to a young forest. A section of boardwalk, with a short spur leading to an open pond, then takes you past a Cattail marsh and through a swamp with Marsh Marigolds and Skunk Cabbage as well as the summer-flowering Lizard's Tail *(Saururus cernuus).* After passing through a thicket of Smooth Alder *(Alnus serrulata),* the trail enters another wooded area, this one largely populated by heaths in the understory: Fetterbush, Mountain

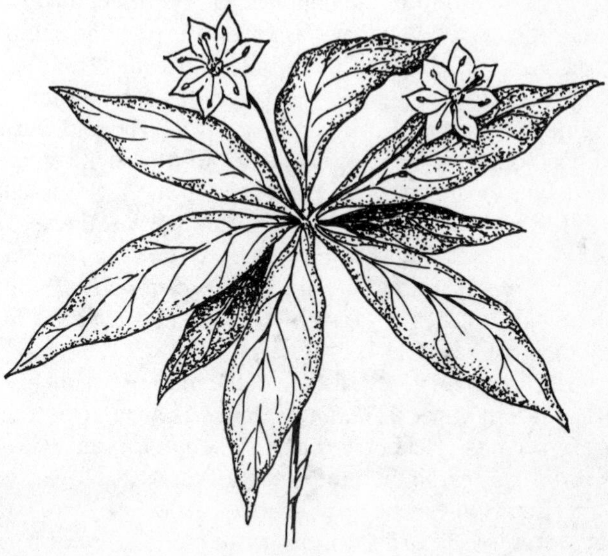

STARFLOWER
Trientalis borealis

Laurel *(Kalmia latifolia),* Highbush Blueberry *(Vaccinium corymbosum),* and Squaw Huckleberry *(V. stamineum).* The last half of the loop consists of a return trip through the swamp and forest.

SANDY HOOK

The superb beaches along New Jersey's 126-mile oceanfront have made it one of the nation's favorite playgrounds, but the development that has followed its growing popularity has had adverse effects upon the land. Only in a few instances has it been possible to preserve the fragile shore with its plant communities reasonably intact.

One such place is Sandy Hook, a narrow sand spit that extends from the Atlantic Highlands into Lower New York Bay, and one explanation for its present condition may be that it has long been under state or federal guardianship. Its strategic importance to the defense of New York City was recognized three centuries ago, and a National Historic Area has been established around the now inactive Fort Hancock. America's oldest original lighthouse, built in 1764, is still in operation here.

Now a component of the Gateway National Recreation Area, Sandy Hook is eminently accessible to residents of greater New York. It is reached via S.R. 36 at Highlands, which is only 13 miles from exit 117 on the Garden State Parkway.

Sandy Hook comprises a wide range of habitats. These include sand dunes, bay shores, fresh and salt water marshes, thickets, and a remarkable forest of American Holly trees *(Ilex opaca),* some of which are more than 300 years old.

Two miles inside the entrance, at Spermaceti Cove, there is a visitor center housed in a building that was erected in 1848 as a lifesaving station. Near this structure is the head of the 1-mile Old Dune Trail.

One feature of seaside vegetation that can be somewhat disconcerting is an abundance of Poison Ivy *(Rhus radicans).* The Old Dune Trail is no exception, and although it is easy enough to avoid contact, this requires looking up as well as down, for much of it grows here in the form of tall shrubs.

The trail commences through head-high Groundsel Trees *(Baccharis halimifolia),* followed by Northern Bayberry *(Myrica pensylvanica),* Winged Sumac *(Rhus copallina),* and Salt Spray Rose *(Rosa rugosa).* Where these shrubs recede, there are patches of Prickly Pear Cactus *(Opuntia humifusa),* a significant member of Sandy Hook's flora.

A wide opening reveals one of the most prevalent herbaceous flowers, Starry False Solomon's-seal *(Smilacina stellata).* Around the edges are other deciduous trees such as Hackberry *(Celtis occidentalis)* and Wild Black Cherry *(Prunus serotina).* The trail then reenters the woods in a forest dominated by handsome American Hollies (the most venerable specimens are in a section now set aside as a preserve in order to protect them from excessive foot traffic).

On the left is a short U-shaped birding loop trail. This is replete with plants bearing succulent fruits that are attractive to birds (which were, in fact, responsible for bringing most of the plants here in the first place). There are Greenbriers (*Smilax* spp.), Serviceberry *(Amelanchier canadensis),* Highbush Blueberry *(Vaccinium corymbosum),* Sumacs (*Rhus* spp.), and Virginia Creeper *(Parthenocissus quinquefolia),* to mention a few.

Emerging from the woods, the path runs close behind the dunes, where Beach Plums *(Prunus maritima)* are very numerous. These straggly shrubs bear copious white blossoms, followed by delicious bright purple, perfectly round fruits. Also in evidence here—although not at all widespread on Sandy Hook—are the low cushions of Beach Heather *(Hudsonia tomentosa).*

Where a sand road intersects the trail, go straight ahead and climb to a low platform for a view of a small freshwater pond. In the surrounding stands of tall Reed Grass *(Phragmites australis),* look for the big pink blooms of Swamp Rose Mallow *(Hibiscus palustris).*

Shortly beyond this point the trail turns to the right, after which it is unmarked; the Atlantic Ocean is directly in front, and in returning to the visitor center you have the choice of beachcombing or walking along the foredunes. The latter route is dominated by Beach Grass *(Ammophila breviligulata),* Seaside Goldenrod *(Solidago sempervirens),* and Dusty Miller *(Artemisia stelleriana).* Other dune plants you are likely to see are Cocklebur *(Xanthium echinatum),* Saltwort *(Salsola kali),* Seaside Spurge *(Euphorbia polygonifolia),* Sea Rocket *(Cakile edentula),* and Trailing Wild Bean *(Strophostyles helvola).*

4

The New Jersey Pine Barrens

UNTIL THEY HAVE seen them, many people simply do not believe that the Pine Barrens exist. They unfold a highway map, but the name does not appear anywhere. This is, after all, highly industrialized New Jersey, our most densely populated state, with only 7,836 square miles to accommodate its 7,364,000 people. Where can there possibly be room for a 2,000-square-mile tract of primitive woods and swamps, only half of it inhabited and sparsely at that?

Well, draw a line just a few miles back from the seashore to connect Asbury Park with Cape May Court House; this gives you the eastern boundary and the approximate upper and lower limits. The western edge is roughly parallel to the Delaware River, and for the most part less than 20 miles from it. The region within this outline is crisscrossed by highways, which provide access to it for the relatively few residents and for the curious tourists but exist primarily to handle the traffic between large urban and recreational centers just outside its borders. The perimeter of the Pine Barrens is within 40 air miles of New York's Times Square, less than 20 miles from Independence Hall in Philadelphia, and a mere 8 miles from the boardwalk and casinos of Atlantic City.

There are a number of reasons why this immense piece of real estate, situated in the heart of the eastern megalopolis, has escaped large-scale economic development. The earliest of these is explained by the name by which the area is still known. The term "barrens" had its origin in the disappointment experienced by settlers who were forced by overcrowding out of the lush Delaware River valley. Reconnoitering the coastal plain to the east, they encountered extensive pine forests growing in sandy soil that they perceived as being relatively infertile and therefore holding little promise of productivity as farmland—in a word, barren. Their assessment was correct enough given their specific requirements, and remained valid for succeeding generations with similar aspirations.

With the passage of time, other opportunities loomed, only to be extinguished by other obstacles. In the 1960s, plans for exploitation culminated in the vision of a supersonic jetport, to be the largest airport in the world by far, together with a new ultramodern city; but so far the Pine Barrens have been spared.

Despite the history of frustration, it would be a gross mistake for us objectively to characterize the Pine Barrens as unproductive. Although much of it looks bone dry—as do all sandy landscapes—just beneath the surface there lies an enormous reservoir of pure water (in a scheme to tap it for supplying the city of Philadelphia, financier Joseph Wharton once bought up immense acreage, but was thwarted when New Jersey refused to allow the water to be exported). Its flat topography notwithstanding, it is far from being a monotonous expanse; in fact, no fewer than 13 distinct phytogeographic formations have been identified within its boundaries.

Hints of this diversity can be seen in the modest enterprises that succeeded, at least for a while, where agriculture had failed. The vast cover of pine forest was quickly turned to advantage; although the dominant Pitch Pine *(Pinus rigida)* furnished inferior timber, it was the species available in great quantity. It also provided crude turpentine, fat pine knots for fuel and illumination, and charcoal for the bog-iron furnaces that supplied munitions for the American Revolution. In the swamps, Atlantic White Cedar trees *(Chamaecyparis thyoides)* were harvested for their extremely durable wood, much of it for splitting into shakes. Native species of Blueberries *(Vaccinium)* and Huckleberries *(Gaylussacia)* throve in the more acid soils, and this led to the cultivation of improved Blueberry strains as an important industry. Flooded peat bogs now produce another indigenous heath, the Cranberry *(V. macrocarpon),* on a commercial scale. But these last two represent the only surviving enterprises of any significance. The gathering of Sphagnum moss for nurseries, medicinal plants for drug manufacturers, and greens and cones and wildflowers for decoration has declined, and the glass works and paper mills that also depended upon natural resources have all but disappeared.

The botanist in us, though, will apprehend the variety and abundance of the Pine Barrens flora immediately, and this can be illustrated very well by my own experience. My introduction to the region came early one July as I parked my car alongside S.R. 530 below Browns Mills and stepped out onto the shoulder. Instantly the realization was upon me that I was walking on orchids, two kinds, in fact: Rose Pogonia *(Pogonia ophioglossoides)* and Grass Pink *(Calopogon tuberosus).* Little round heads of brilliant orange turned out to be a Milkwort, *Polygala lutea,* and these were surrounded by a great many dull rose-purple

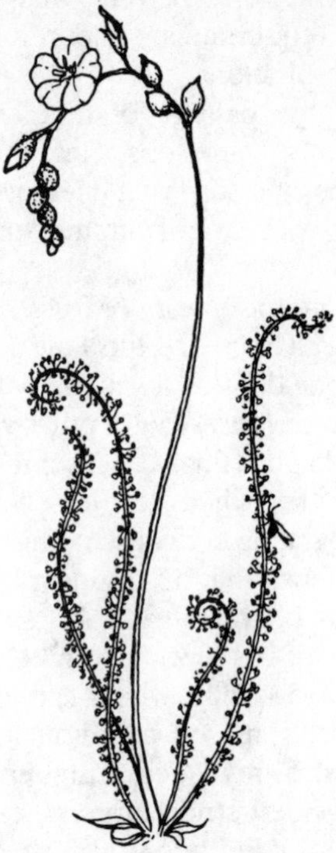

THREAD-LEAVED SUNDEW
Drosera filiformis

Cross-leaved Milkworts *(P. cruciata).* Adding white to the scene were diffuse clumps of Pine Barren Sandwort *(Arenaria caroliniana)* and tall, slender spikes of Colic Root *(Aletris farinosa).* Tracing wiry vines toward their ends revealed the nodding blossoms of Cranberries *(Vaccinium macrocarpon).* This indicated wetness nearby, and a short search led to the edge of a boggy spot with Pitcher Plants *(Sarracenia purpurea),* yellow Bladderworts (*Utricularia* sp.), and Thread-leaved Sundew *(Drosera filiformis).* At the time most of these species were new to me, but accompanying them were some old acquaintances, such as Sheep Laurel *(Kalmia angustifolia),* Cow Wheat *(Melampyrum lineare),* and hundreds of blooms of Rabbitfoot Clover *(Trifolium arvense)* massed to form a border of delicate, misty mauve along the margin of the berm. Finally, some spikes of reddish snapdragonlike flowers caught my eye, and these keyed out to *Schwalbea americana,* or Chaffseed—yet another first.

THE PLAINS

Certainly the most extraordinary part of the Pine Barrens is that known simply as "the Plains," a sandy 12,000-acre expanse blanketed by a mixture of stunted oaks and pines. These gnarled pygmy trees are almost uniformly 4 feet in height, which makes it possible to stand among them and see over their tops in all directions.

The reasons for this phenomenon have eluded experts ever since Gifford Pinchot, America's preeminent forester, first tried to probe the mystery. The poverty of the soil, repeated burnings, exposure to harsh winds, and genetics have all been suggested as contributing factors.

The principal tree species are Pitch Pine and Blackjack Oak *(Quercus marilandica),* with Post and Bear Oaks (*Q. stellata* and *Q. ilicifolia*) much less common. Scattered throughout are various shrubby heaths, among them Sheep Laurel *(Kalmia angustifolia),* Black Huckleberry *(Gaylussacia baccata),* and Sand Myrtle *(Leiophyllum buxifolium).* Low-growing species are more the rule, however, with Bearberry *(Arctostaphylos uva-ursi)* plentiful and Trailing Arbutus *(Epigaea repens)* especially abundant. Other plant families are represented by Goat's Rue *(Tephrosia virginiana),* Downy False Heather *(Hudsonia ericoides),* and Golden Aster *(Chrysopsis mariana).* The region's commonest fern is Bracken *(Pteridium aquilinum),* and here it is interspersed with silvery tufts of Reindeer Lichen *(Cladonia rangiferina).*

An attractive plant seldom seen elsewhere is Pyxie Moss *(Pyxidanthera barbulata),* not a moss at all but a relative of the more southern Galax and of the boreal Diapensia. Pyxie forms dense mats of crowded evergreen leaves, which in some years assume a deep red tint. These are all but concealed in the spring by multitudes of sessile white flowers.

The most unusual plant of the Plains is Broom Crowberry *(Corema conradii),* which comes into flower in early April. This small shrub is a relict species that at one time probably was distributed continuously along the Atlantic seaboard but now occurs in the United States only at widely separated stations. It is

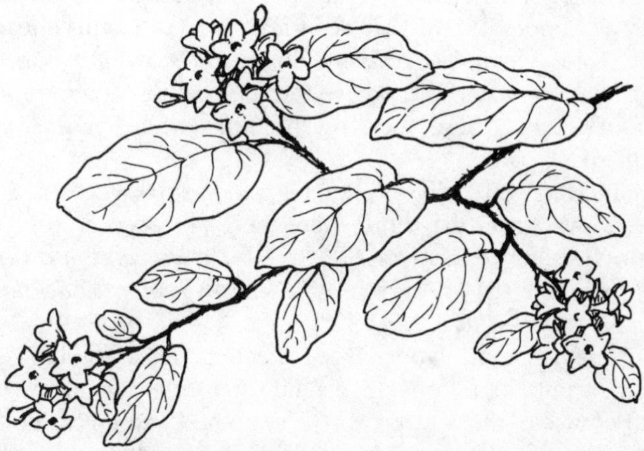

TRAILING ARBUTUS
Epigaea repens

dioecious, and both the pistils and stamens, including the anthers, are purple; but since they lack petals, they are quite inconspicuous.

The lower section of the Plains lies west of S.R. 539 beginning a few miles north of the Garden State Parkway. This is separated by the east branch of the Oswego (or Wading) River from the upper section, which is bisected by S.R. 72 westward from its junction with S.R. 539. Setting out on foot, you will see vestiges of old roads meandering about, leading to no destination of particular interest to the visitor but affording a nice random system of trails for walking over the Plains.

PAKIM POND

After a period of wandering (and wondering) about the Plains, it might be a refreshing change to visit Pakim Pond in Lebanon State Forest. This placid, green-bordered body of water once served as a reservoir in which water was stored for flooding a cranberry bog (the name is derived from the Leni Lenape Indian word for cranberry). The well-marked entrance is on the north side of S.R. 72 about 3 miles east of the junction with S.R. 70.

Pakim Pond is one of many access points for hikers who may wish to use the Batona (BAck-TO-NAture) Trail, which traverses 41 miles of easily negotiated wilderness in the Pine Barrens; but for those with less time there is an excellent 1-mile self-guiding nature trail. This starts out at the small pavilion near the bathing beach and circles the pond.

Among the woody plants at the beginning of the trail are Gallberry, or Inkberry Holly *(Ilex glabra),* which bears blue-black fruits; Maleberry *(Lyonia ligustrina);* and the lower-growing Staggerbush *(L. mariana).* The edges of the pond have a variety of insectivorous plants. The largest are the Pitcher Plants, but three kinds of Sundews have also been found here: Spatulate-leaved, Round-leaved, and Thread-leaved (*Drosera intermedia, D. rotundifolia,* and *D. filiformis*). In shallow water are the yellow flowers of Common Bladderwort *(Utricularia vulgaris);* here too are the little white buttons of Pipewort (*Eriocaulon* sp.).

One should back off every so often for a look at the vegetation in the drier sandy habitat away from the shore. Here there are deep crimson-pink Sheep Laurel, the rich yellow, ephemeral Frostweed *(Helianthemum canadense),* and the dramatic white heads of Turkeybeard *(Xerophyllum asphodeloides)* standing above clumps of grasslike foliage.

Where a small stream, Cooper Branch, enters Pakim Pond from an Atlantic White Cedar swamp, look for Golden Club *(Orontium aquaticum),* also called Never Wet because of the way its broad leaves shed water. Farther on there are Red Maples and a few other deciduous trees, with an understory dominated by Sweet Pepperbush *(Clethra alnifolia)* and Highbush Blueberry *(Vaccinium corymbosum)*—and, of course, the symbol of the region, Pitch Pine.

CEDAR SWAMPS

As incongruous as it may sound, a good deal of the Pine Barrens consists of Atlantic White Cedar swamps. These no longer represent an important economic asset, as they did when the trees were being harvested for their light, durable wood, and the giant specimens are gone, but none of the fascination that we find in dark, mysterious, primeval forests has been lost.

The sturdy cedars with their straight trunks and buttressed bases form close, and often nearly pure stands. The pattern is sometimes broken by the small-flowered Sweet Bay *(Magnolia virginica),* Sour Gum *(Nyssa sylvatica),* or Red Maple. Shrubs may be expected to include Sweet Pepperbush, Maleberry, Swamp Azalea *(Rhododendron viscosum),* Virginia Willow *(Itea virginica),* and a variety of Huckleberries (*Gaylussacia* spp.), while Greenbriers such as *Smilax walteri,* which has red berries, and *S. laurifolia* are among the vines. Cinnamon Ferns *(Osmunda cinnamomea)* build up mounded tufts that provide surprisingly solid footing; the other principal fern is Netted Chain Fern *(Woodwardia areolata).* And wherever the ground is not submerged beneath the brandy-colored water, there is a thick carpeting of Sphagnum moss.

Even filtered sunlight is sparse here, and so are flowering herbs, but the few are well worth the search. At the top of most want lists is *Arethusa bulbosa,* called by some Dragon's Mouth. This striking little orchid, which lacks any sign of a leaf at the time of flowering, stands stiffly erect above its mossy bed displaying a wide gold and purple crested lip. A frequent companion is Rose Pogonia, or Snake Mouth *(Pogonia ophioglossoides),* also an orchid but one of more graceful carriage, bearing a solitary flower of exquisite light pink.

Showier than either of these is Swamp Pink *(Helonias bullata),* a member of the Lily family. From a rosette of evergreen leaves, a single hollow stem rises to a height of a foot or more. This is surmounted by an egg-shaped raceme of densely crowded flowers, pink with blue anthers and exuding a spicy fragrance.

Even the largest of these swamps is not identified by signs, as we might expect. You must seek them out either by making inquiries or by looking for concentrations of the unique flattened foliage of the cedars as you drive along the roads. Nor are there paths or trails into them. You enter wherever you can—preferably near one of the branches, as advertised by a small bridge—the only requirement being a foundation of tree roots or fern hummocks beneath the blanket of Sphagnum moss substantial enough to support your weight. Since this inevitably leads you on an erratic course, you must maintain your orientation by means of a compass or otherwise, for you quickly lose sight of your starting point in the dense growth.

A number of white cedar swamps are situated in the western part of Lebanon State Forest, and directions can be obtained at the headquarters and visitor center. This can be reached from S.R. 72 by taking the Pakim Pond entrance road, or from S.R. 70 just east of its junction with S.R. 72 (this intersection, known as Four Mile Circle, is convenient to use as a reference point in directions to many locations in the Pine Barrens).

You can, of course, strike out on your own instead. For example, if you drive east from the visitor center for three-quarters of a mile, then turn left, you will cross Shinn's Branch. The next left soon comes to Cooper's Branch, and shortly beyond that McDonald's Branch. The first two are in typical cedar swamps, whereas McDonald's flows through a more open Red Maple swamp at this point, with Gray Birches *(Betula populifolia)* on higher ground. Leatherleaf *(Chamaedaphne calyculata)* fringes the stream, and Spatterdock *(Nuphar variegatum)* is in the open water.

An exceptional example of a cedar swamp can be reached from S.R. 70 by taking the road marked "Mt. Misery" for 1 mile, taking the left-hand fork for 2 more miles, going right on a sand-gravel road for a third of a mile, and parking near the bridge. On the right is a heavily wooded cedar swamp, but across the road the scene is very different. Cleared at some time in the past of all its trees, this part of the swamp was converted into a bog (very likely for the growing of cranberries), and there is something left of a rotting corduroy road on which some progress can be made into the squishy interior. Hundreds of small, emerald green cedars signal the start of its long, slow return to swampland, but for the time being it is bathed in sunlight and wildflowers are flourishing. One especially attractive plant, although elusive in its blooming habits, is Marsh St. John's-wort *(Triadenum virginicum)*; its flowers are pink and feature conspicuous orange glands. The closely related Canadian St. John's-wort *(Hypericum canadense)* has much smaller yellow flowers. Others include Meadow Beauty *(Rhexia virginica)*, Yellow-eyed Grass *(Xyris carolina)*, and the semi-woody Swamp Loosestrife *(Decodon verticillatus)*.

MARTHA'S FURNACE

If "Martha's Furnace" does not sound like a good place to botanize, prepare for a surprise, for here you will find what is perhaps the richest flora the Pine Barrens have to offer. The site is now within Wharton State Forest, and the little town of Martha and its 1793 bog-iron smelter are gone without a trace except for some shallow cellar holes and a few Catalpa trees.

Canoeists come here on the silently flowing Wading River, which like most of the other streams in the area has taken on the color of strong tea from infusions of tannin and minerals. Martha's Furnace can easily be reached overland, however, by driving south from Chatsworth on S.R. 563 past blueberry farms and commercial cranberry bogs. At a fork, bear left on the 563 spur, which may appear on newer maps as S.R. 679. The 1-mile road to Martha's Furnace branches off to the northeast just after this intersection. This sand road may have depressions filled with soft sand or standing water, and unless you have a four-wheel-drive vehicle it might be advisable to walk in from here.

The most important floral feature of the wet savanna at Martha's Furnace is an exceptionally large stand of the rare Bog Asphodel *(Narthecium americanum).* These plants bear racemes of flowers that are yellow in all their parts, including even the woolly hairs that clothe the long, straight filaments. Accompanying them is another liliaceous plant—the closely related, white-flowered False Asphodel *(Tofieldia racemosa),* as well as the very dissimilar Golden Crest *(Lophiola americana).* The open, branched inflorescence of Golden Crest is covered by a dense, felty, white tomentum that completely masks the green color. The perianth consists of six segments that spread out to reveal reddish brown inner surfaces and thick tufts of golden yellow hairs at their bases. All through this meadowlike area are the white "hat pins" of Pipewort *(Eriocaulon decangulare).*

In drier sections there are such flowers as Orange Milkwort, Pine Barren Sandwort, Turkeybeard, Sundrops *(Oenothera perennis),* and *Sabatia difformis,* a coastal species with pure white petals lacking the central starlike blaze that characterizes many of the so-called Marsh Pinks.

In the other direction, nearer the open water, look for three of our most beautiful native orchids, Arethusa, Grass Pink, and Rose Pogonia. Large open, muddy spaces are strewn with the yellow blossoms of Horned Bladderwort *(Utricularia cornuta)* and the smaller Zigzag Bladderwort *(U. subulata),* interrupted in many places by clumps of Pitcher Plant. As might be expected in this boggy environment, several Sundews are also present, with Round-leaved and Spatulate-leaved surpassed in numbers by the unusual Thread-leaved species, whose bladeless leaves are invested from top to bottom with countless glistening glandular hairs.

In the early 1800s the New Jersey Pine Barrens attracted worldwide attention when Frederick Pursh discovered the Curly Grass Fern *(Schizaea pusilla)* at Quaker Bridge, not far from here. This rare fern occurs in a number of other sites in the Pine Barrens, but Martha's Furnace is a good place to see it. Here it occupies small hummocks in the shallows along the edge of the river, and in searching for it one should take care to walk only in the water to preclude the possibility of stepping on it. As the drawing on the next page indicates, the twisted sterile leaves account for the common name and indeed could easily be mistaken for those of a grass. Doubtless this unique little fern owes its survival in large part to its insignificance (it is only 3 or 4 inches high), which deprives it of any appeal as a garden ornamental.

ATSION LAKE

Driving in the Pine Barrens, especially through state forest lands, you cannot help being aware of the variety in the roadside vegetation. Some of it puts on

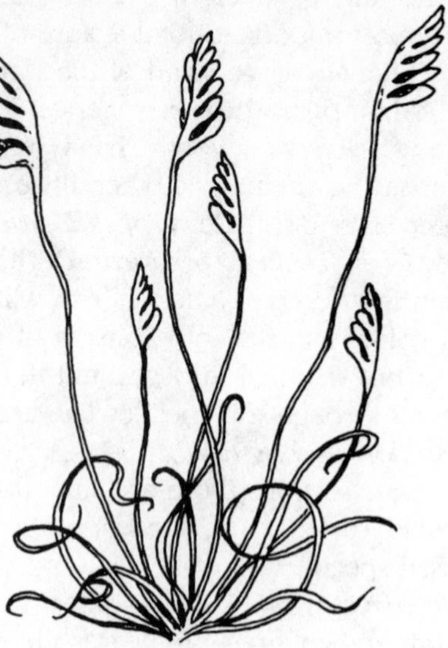

CURLY GRASS FERN
Schizaea pusilla

displays that can hardly be ignored: myriad pink clusters of Mountain Laurel *(Kalmia latifolia),* sheets of Water Lilies *(Nymphaea odorata)* on quiet ponds, and drifts made up of the little Wild Pansy *(Viola rafinesquii)* multiplied a thousand times.

Obviously, much more can be seen by stopping, as has already been suggested in recitals of swamp and bog flora. In the sandy pine woods, you will find shrubs like Fetterbush *(Leucothoe racemosa),* as well as such diverse small herbs as Birdfoot Violet *(Viola pedata),* Nodding Ladies' Tresses *(Spiranthes cernua),* Tawny Cotton Grass *(Eriophorum virginicum),* St. Peter's-wort *(Hypericum stans),* and the extremely variable Ipecac Spurge *(Euphorbia ipecacuanhae).*

One particular locale that should not be overlooked is Atsion, on U.S. 206 about 10 miles north of Hammonton, where a public recreation area for swimming and picnicking has been provided at Atsion Lake.

For the best botanizing in this vicinity, look for low, moist places along the shoulders of the roads and beside the old railroad tracks. Representing the monocots here we have Redroot *(Lacnanthes caroliniana); Platanthera blephariglottis* var. *conspicua,* which is a White Fringed Orchid distinguished by an exceptionally long spur; and Slender Blue Flag *(Iris prismatica)* in the ditches. Other plants range from the vivid Virginia Meadow Beauty to the modest Nuttall's Lobelia *(Lobelia nuttallii)* to the inconspicuous but interesting semiaquatic herb with fine pinnately divided foliage known as Mermaid Weed, or *Proserpinaca pectinata.* The wildflower prize would certainly have to go to the Pine Barren Gentian (*Gentiana autumnalis,* or *G. porphyrio* of some authors). This rare species has very narrow leaves and a flower with spreading lobes, brilliant blue inside with greenish spots in the throat.

PINE BARREN GENTIAN
Gentiana autumnalis

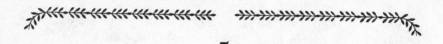

5

The Southern Appalachians

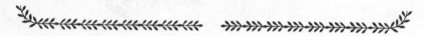

THE MOST DIFFICULT aspect of describing what we call the "southern" Appalachians is trying to delineate their boundaries with any degree of precision. According to strict interpretation, they are said to begin in southern Virginia, straddle the North Carolina–Tennessee line, and terminate in northern Georgia. This will suit our purposes well enough, but now and then we will permit an irresistible botanical area to entice us into places like West Virginia and South Carolina.

The rounded, gently undulating ridges that make up these mountains tell us of their great age, which far surpasses that of the Alps or the Rockies, but also tend to obscure the existence among them of the highest summits in eastern North America. The blue haze that separates their receding planes, making them look like a series of theatrical scenery flats, is the product of the luxuriant vegetation in which they are clothed.

Closer examination reveals their most engaging attribute—a remarkable mosaic of habitats comprising nearly every conceivable forest plant community and partaking generously of both northern and southern species.

At the very highest level there are "islands" of spruce and fir, extensions of the coniferous forests much more common in the northern reaches of the United States. At slightly lower elevations these are joined by a rich mixture of hardwoods, heath shrubs, and forbs. The most diverse vegetation, as well as fine examples of virgin timber, is found in the cove forests, where sheltered ravines with tumbling streams are home to an astonishing array of tree species and spectacular displays of wildflowers. Unique to the southern Appalachians and baffling as to origin are the mountain balds, treeless highlands in a region where there is no treeline. Each of these habitats is subject in turn to various environmental influences, and the result is an extraordinarily complex flora.

The southern Appalachians were really opened up to botanical exploration by William Bartram shortly after the beginning of the American Revolution, but probably the best-known story of these early endeavors is that of the discovery of *Shortia galacifolia* (now known as Oconee Bells) by André Michaux around 1788 in the western Carolinas. He collected a single specimen without petals (it had bloomed earlier) and eventually placed it in his Paris herbarium. When the famous American botanist Asa Gray came across the sheet in 1839, he deemed it a new genus and named it for Dr. Charles Short of Kentucky. The plant was not seen again until 1877, when it was rediscovered by 17-year-old George Hyams. A flowering specimen was sent to Dr. Gray at Harvard, but although he searched diligently throughout the rest of his life, he never saw *Shortia* blooming in the wild.

Much of *Shortia*'s original habitat has been lost, and most of the remaining sites are in obscure locations. Fortunately, however, a great many plants were rescued from imminent destruction and have been transplanted extensively. As a result, thriving populations of this exquisite flower can now be enjoyed in many public and private gardens without fear of adverse impact upon the species.

We can scarcely imagine the difficulties that met the pioneer botanists' efforts to penetrate this hostile wilderness. Today we are able to explore much

OCONEE BELLS
Shortia galacifolia

of it in comfort, at our convenience, and at no cost, thanks in large measure to the existence of the Blue Ridge Parkway and Great Smoky Mountains National Park. The following sections present an overview of these splendid facilities. Examples of their most salient plant communities—spruce-fir forests, cove hardwoods, and mountain balds—are described in detail in later chapters.

BLUE RIDGE PARKWAY

From the lower end of Shenandoah National Park near Waynesboro, Va., the ribbonlike Blue Ridge Parkway follows the eastern escarpment of the southern Appalachians for 469 miles through North Carolina to the Oconaluftee gateway to the Great Smoky Mountains National Park, near Cherokee. Pinpointing any location along its route is facilitated by milepost markers, which are numbered starting at the Virginia terminus and are located on the western shoulder.

The Blue Ridge Parkway is a national park also, flanked over much of its length by unspoiled woods that open out frequently at overlooks to reveal distant views, but by its very nature it is a *landscaped* national park, and this concept permits a greater degree of management than would be tolerated elsewhere in the system. The National Park Service conscientiously seeks, and tries to follow, the advice of competent botanists, but inevitably its highway maintenance operations sometimes prove inimical to the natural regeneration of roadside vegetation. For this reason the reported presence of a particular species at a given location must not be taken as guaranteed.

FLOWERING DOGWOOD
Cornus florida

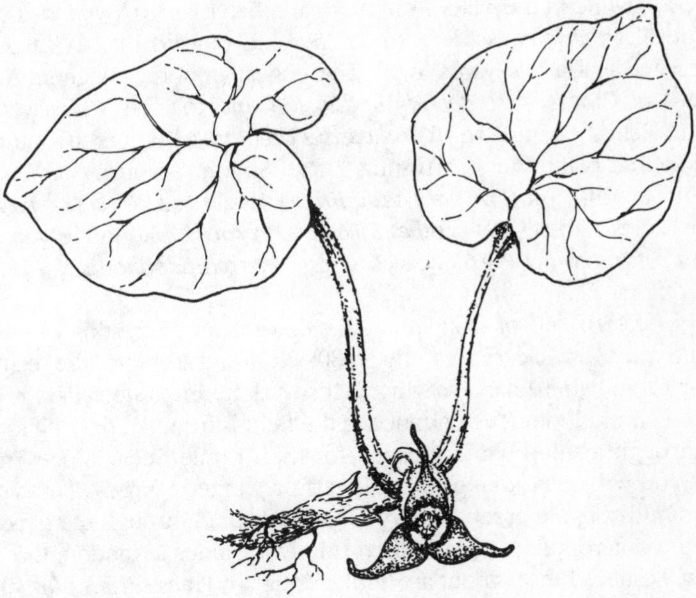

WILD GINGER
Asarum canadense

Against the backdrop of oak-hickory forest, which is the dominant cover, the plant life along the parkway increases in richness and variety as we progress southward. This is due largely to the fact that, in general, the North Carolina portion receives heavier annual precipitation than the Virginia segment. It also contains more high-altitude areas, and these foster whole groups of plants that are better suited to cold climatic conditions.

Smart View, south of Roanoke at mile 154.5, has a loop trail that affords a pleasant introduction to the parkway flora in Virginia. The woods here consist mainly of Oaks—Red, Scarlet, Black, Chestnut, and White among them. Flowering Dogwood *(Cornus florida)* is the chief understory tree and is especially attractive in early May when seen in proximity to the rustic log cabin and split-rail fence. Pinxter Flower *(Rhododendron periclymenoides)* and Flame Azalea *(R. calendulaceum)* are two of the more colorful shrubs. Wildflowers blooming at the same time along the trail include Large-flowered Trillium *(Trillium grandiflorum),* Wild Columbine *(Aquilegia canadensis),* Showy Orchis *(Galearis spectabilis),* Fire Pink *(Silene virginica),* and Wild Ginger *(Asarum canadense).* Occasional plants of the fleshy Pennywort *(Obolaria virginica)* may also be sighted emerging from the leaf litter.

Little Glade Pond, at mile 230.1, provides an unusual habitat for the parkway, and is a delightful spot for a picnic lunch. The margins of the pond have Tag

Alder *(Alnus serrulata),* Swamp Rose *(Rosa palustris),* and Steeplebush *(Spiraea tomentosa),* while in the water itself there are Water Lilies *(Nymphaea odorata)* and Blue Flags *(Iris versicolor),* to say nothing of wild mallards that dabble unconcerned just a few yards away. The edges support Sphagnum Moss and Running Pine Clubmoss *(Lycopodium flabelliforme)* together with several ferns, including Netted Chain Fern *(Woodwardia areolata).* But it is the multitude of small-flowered plants that is arresting: Dwarf St. John's-wort *(Hypericum mutilum),* Purple Milkwort *(Polygala sanguinea),* Bugleweed *(Lycopus virginicus),* Slender Ladies' Tresses *(Spiranthes lacera* var. *gracilis),* Round-leaved Sundew *(Drosera rotundifolia),* Purple-leaved Willow Herb *(Epilobium coloratum),* and several others.

At the eastern end of Doughton Park there is an interesting example of a mountain meadow cleared in the 1880s. It is adjacent to the century-old Brinegar cabin, where handweaving as it was done in pioneer days is demonstrated on an old loom. The entrance to the field is from the parking lot at mile 238.5, through a stile behind the sign for the 11-mile Cedar Ridge Trail.

First you will come to a patch of Dwarf or Winged Sumac *(Rhus copallina)* with pyramids of pale green flowers. Continue uphill through the grassy clearings to the outcroppings of rock, then simply meander at random through the meadow. Scattered throughout are Table Mountain Pines *(Pinus pungens).* The rocks have Hay-scented Fern *(Dennstaedtia punctilobula)* and Southern Harebell *(Campanula divaricata)* in the crevices and Pineweed *(Hypericum gentianoides)* around the base. Most of the flowers here come into their own during summer and fall, and the majority, predictably, are European species that owe their success in the New World to their ability to make the most of such open situations. Tall enough to stand above the grasses, they form a sea of brilliant color.

The Linville Gorge Wilderness Area is ruggedly beautiful country with rather strenuous hiking trails, but an easy 1-mile trail to Linville Falls has been provided from a parking lot just south of the parkway at mile 316.3. This path ascends through a climax forest notable for its large Carolina Hemlocks *(Tsuga caroliniana),* branches out to three very different viewpoints of the falls, and ends with a panorama of the gorge. Of the many interesting trailside plants, perhaps the most unusual is Turkeybeard *(Xerophyllum asphodeloides).*

No motorist traveling along the parkway in the vicinity of milepost 417 will miss seeing the gray dome of Looking Glass Rock, a majestic granite monadnock rising 400 feet above the surrounding woodland to an altitude of 3,969 feet. It gets its name from the sheen on its smooth surface when wet with rain or ice. Forbidding as it may appear, it is entirely feasible to hike to the top and sit down to a trail lunch on the very brow that faces the parkway. The trick is to get to the other end of the rock and walk up its sloping back. This is done by turning off onto U.S. 276 south, then making a right after 9.5 miles on Forest Service Road 475 at a sign for the State Fish Hatchery; parking is a half mile in on the right.

The trail ascends 1,370 feet in 3 miles, but thanks to many switchbacks it is quite gradual. At the start it passes through a grove of Eastern Hemlock *(Tsuga canadensis),* but in the upper portion there are Carolina Hemlocks, recognizable even at a distance by their bright green color. The summit is just as rounded and smooth as it appears from below and the view is utterly breathtaking. A particularly lovely sight comes in late May when the Fringe Trees *(Chionanthus virginicus)* on the top form a white and green border at the forest edge.

A shorter deviation from the parkway at mile 423.3 can also be very worthwhile. In early May, you might wish to drive down N.C. 215 in the direction of Rosman for a few miles to enjoy the Silverbell trees *(Halesia carolina)* and the Pink-shell Azaleas *(Rhododendron vaseyi);* the latter is a beautiful species whose range is limited to the mountains of western North Carolina. If you are here in September, park in one of the wide spaces on the left shortly after leaving the parkway and inspect the banks across the road. The seeps have drifts of Grass-of-Parnassus *(Parnassia asarifolia)* intermixed with blue Closed Gentian *(Gentiana clausa)* and Pink Turtlehead *(Chelone obliqua)* and stands of American Burnet *(Sanguisorba canadensis),* a plant found only in the mountains this far south. Also infrequent in this region is the delicate Pale Corydalis *(Corydalis sempervirens),* which you may see growing higher up in dry rock crevices.

Back on the parkway, you will be tempted to stop at several more of these rocky seeps. Near Devil's Courthouse Tunnel (mile 422.1) they are blanketed with *Hypericum buckleyi,* a prostrate St. John's-wort found only in this and a few adjoining counties. Walking along the wet rock faces farther south, you will find three other species—*H. graveolens, H. densiflorum,* and *H. mutilum*—as well as False Asphodel *(Tofieldia racemosa* var. *glutinosa),* Club-spur Orchid *(Platanthera clavellata),* Cowbane *(Oxypolis rigidior),* Mountain Cynthia *(Krigia montana),* Michaux's Saxifrage *(Saxifraga michauxii),* and sheets of Thyme-leaved Bluets *(Houstonia serpyllifolia).*

Several flowering plants occur in conspicuous but infrequent patches along the roadway, some examples being the following:

Species	**Mile**
Blazing Star *(Liatris spicata)*	369.9
Cardinal Flower *(Lobelia cardinalis)*	North of 378
Small-flowered Phacelia *(Phacelia dubia)*	Craven Gap, near 378
Obedient Plant *(Dracocephalum virginianum)*	South of N.C. 151 (405.5)
Indian Paintbrush *(Castilleja coccinea)*	411.3
Wild Larkspur *(Delphinium tricorne)*	440.9

One of the glories of these mountain ridges is the Serviceberry tree *(Amelanchier arborea),* which is sometimes spelled "Sarvisberry" but pronounced that way in any case. When in blossom during early May, these trees appear on distant slopes as puffs of a strange off-white hue; at close range this is seen to be a combination of white petals and taffy-colored young leaves. There are a number of explanations for the colloquial name, one of which is that old-time mountain folk noticed that they bloomed at the time when the ground had thawed enough to finally permit burial, and funeral services, for those who had passed away during the winter.

GREAT SMOKY MOUNTAINS NATIONAL PARK

Compared to many of our other national parks, with their blatant natural features—canyons and geysers, dunes and glaciers, caverns and volcanoes—Great Smoky Mountains National Park seems serene and modest under its cloak of verdure and swathed in the mists that gave it the name. The magnificent mantle of vegetation, though, is not a mark of humility but a badge of pride, for it is unsurpassed in the nation for its variety and luxuriance. The statistics are staggering. In Europe and in our own West, whole mountainsides often are blanketed by trees of a single species; in this national park there are more than 100 kinds. The census of flowering plants comes to better than 1,300 species.

The factors that favored these mountains were many, but all revolve around their geographic location. They were spared inundation by seas and scouring by glaciers, they receive copious amounts of rainfall, they lie in a temperate climate where representatives of boreal and subtropical plant families meet at the extremes of their ranges, and timeworn as they are, they still exhibit wide variations in altitude and configuration. And in the more recent past, 800 square miles of this superb country has been designated a national park to be preserved for all time.

This is not to say that the Great Smokies are locked up away from the public; on the contrary, they are visited by more people than any other national park. Situated astride the Tennessee–North Carolina border, the park is bisected from north to south by U.S. 441, which connects Gatlinburg, Tenn., with Cherokee, N.C., and meets the lower end of the Blue Ridge Parkway near the latter town. Just inside the park boundaries are two visitor centers, Sugarlands in the north and Oconaluftee in the south. Except for U.S. 441 and a spur road to Clingman's Dome, the park roads are mainly peripheral; only the Appalachian Trail runs along its long axis.

With moderate speed limits and ample pullouts, the park roads offer opportunities for viewing wildflowers and other vegetation. In addition, they take you to the starting points for the many hiking and self-guiding nature trails, and to the innovative "quiet walkways." These paths, ranging from a short stroll to several miles graded for easy walking, were designed, according to one park ranger, "to get people out of their cars and into the woods," and can be most enjoyable. A half-hour stop at one of these provided a pleasant surprise in the form of large numbers of Dwarf Ginseng *(Panax trifolium)* and Golden Saxifrage *(Chrysosplenium americanum),* neither of which is very plentiful in this area.

The drive from Sugarlands through Little River Gorge is probably the most beautiful in the park, and is a good place to see Cross-vine *(Bignonia capreolata)* and the rare Yellowwood *(Cladrastis lutea),* a handsome tree with white wisterialike blossoms, as well as low-altitude wildflowers. It leads to the Cades Cove Road, which loops through a group of restored pioneer homesteads and churches, and from which two short trails can be taken. One is to Abrams Falls, where Mountain Spleenwort *(Asplenium montanum)* shares the rock faces with the more common Maidenhair Spleenwort *(A. trichomanes).* The other is the self-guiding Cades Cove Vista Nature Trail, for which there is a leaflet explaining uses of plants by the early settlers.

The Roaring Fork Motor Nature Trail is a 2.5-mile loop starting on Airport Road in Gatlinburg. From this an easy trail goes through a hemlock forest to Grotto Falls, where it is possible to walk behind the waterfall. An unusual plant along this path is the showy Fraser Sedge *(Cymophyllus fraseri).*

In the southeastern part of the park is the one-way Round Bottom Road, reached by taking the spur at mile 458.2 on the Blue Ridge Parkway to the Heintooga picnic area. Almost immediately, on the right-hand bank, there are masses of the endemic Rugel's Indian Plantain *(Cacalia rugelia).* This road goes deep into the Smokies, then joins Big Cove Road through the Cherokee Indian Reservation to emerge on U.S. 441 near another pioneer farmstead at Oconaluftee.

6

The Other Canada

STRETCHED LIKE a dark green blanket across Canada from Newfoundland to the Yukon is America's vast boreal coniferous forest, the taiga, the classic "North Woods"—land of spruce and fir, moose and muskeg, lynx and loon. Once it extended southward and covered much of what is now the eastern United States. But that was a time of bitter, unrelenting cold, and as the climate moderated and the continental ice sheet slowly receded, trees that were better adapted to milder temperatures began to gain a foothold on the land at the more southern latitudes.

Wherever these broad-leaved species encountered sudden and frequent episodes of subfreezing weather, however, they were unable to compete successfully with the evergreen conifers, which were inherently equipped to cope with such rigors. These inhospitable areas were large at first but gradually contracted, and today the only places in the South that still are too frigid for any but a predominantly fir-spruce forest are the summits of the highest mountains. Because of the topography of the southern Appalachians, most of these summits are within the boundaries of North Carolina.

Notwithstanding their geographic location, these isolated remnants of boreal landscape are known as "Canadian zone" forests, which is an apt designation since they bear a closer resemblance to the woods a thousand miles to the north than to those just a few thousand feet below on the same mountains.

One difference, though, between northern and southern spruce-fir forests is that Balsam Fir *(Abies balsamea),* which ranges from Labrador to southwestern Virginia, is replaced in the mountains of North Carolina and Tennessee by Fraser Fir *(A. fraseri).* This apparently evolved as a separate species during the long isolation of the southern mountains. The only obvious distinction between the two firs is in the erect, dark purple cones: In Fraser Fir the pointed bracts are reflexed and protrude conspicuously from between the scales, but even this clue may prove elusive for the cones disintegrate on the tree at maturity, leaving only the spikelike central axes of the cones. Both species are superlative Christmas trees, as they retain their needles for a long time after being cut.

Pure stands of Fraser Fir are found only at the highest elevations, with Red Spruce *(Picea rubens)* becoming more dominant lower down on the slopes. This is the same Red Spruce that populates the Appalachians all the way to the St. Lawrence and that in the North is a principal source of pulp for paper manufacture. Its chestnut brown woody cones are pendent and fall from the tree soon after shedding the seeds, but they remain intact and can be gathered in quantities from the ground. Even without them, however, spruces can be recognized by touch, as their needles have sharp points, while those of firs are blunt ended.

In local parlance Fraser Fir is known as "She-Balsam," an allusion to the resin-filled blisters beneath its thin bark, which seem to have suggested the tree's giving milk; by default, it appears, Red Spruce has become "He-Balsam." The exudation of certain firs has the same refractive index as optical glass, and this led to its former use as a cement for joining the elements of compound lenses.

A few deciduous trees are able to become established when clear spaces are opened up, for instance, by a fire or a windfall. Fire Cherry *(Prunus pensylvanica)* is, as its common name would indicate, an early pioneer following a burn. It serves as a "nurse tree," sheltering the seedlings of other species, but cannot itself endure shade and succumbs before reaching any appreciable height.

American Mountain Ash *(Sorbus americana),* Mountain Maple *(Acer spicatum),* and Yellow Birch *(Betula lutea)* also are commonly found in association with the conifers. Anyone who is accustomed to seeing European Mountain Ash or Rowan Tree *(S. aucuparia)* planted as an ornamental will see resemblances in American Mountain Ash, although our native species has long, pointed leaflets and fruits that are more red than orange. The yellow flowers of Mountain Maple make it unique within its genus in that they are arranged in an erect raceme and do not open until summer. Mountain Maple is usually considered a shrub, and it is only in the mountains of Tennessee and North Carolina that it attains tree stature. Yellow Birch has bark that is actually gray (with a yellow tinge to its lustrous satiny sheen) and that peels off in narrow curls. The young twigs have a weaker "wintergreen" flavor than those of Black Birch *(B. lenta).* Seedlings of Yellow Birch often become established on moss-covered logs and send roots down on either side; years later, after the log has rotted away, these roots provide a stiltlike base supporting the tree in midair.

Among the shrubs are two attractive Viburnums. The showiest, and the one that prefers colder habitats, is Hobblebush *(Viburnum alnifolium),* so named because of the tangles formed by its reclining branches, which have a habit of rooting at their tips. It has doilylike cymes of small white flowers surrounded by much larger sterile ones, the latter presumably intended to decoy pollinating insects. The fruits are bright orange-red and turn blue-black at maturity. The other is *V. cassinoides,* known as Witherod or Wild Raisin. This species lacks the big sterile flowers, but as its berries ripen they progress from yellow-green through a beautiful shade of rose pink to a glaucous deep blue, often exhibiting all of these colors simultaneously in a single cluster.

Since there is no time when the tree canopy drops all of its leaves to admit the warming sun, this is not the place to look for masses of spring wildflowers. The herbaceous species that do manage, however, include some of the loveliest. Painted Trillium *(Trillium undulatum),* its frilled petals delicately streaked with crimson, gives an impression of daintiness that masks its ability to withstand the severe climate on the highest summits. The handsome Bluebead Lily *(Clintonia borealis)* puts on a fine show in June of yellow bell-shaped flowers overtopping wide satiny leaves, and another in the fall with its bright blue fruits. The name of the genus honors De Witt Clinton, who was not only the governor of New York who sponsored the building of the Erie Canal but an accomplished naturalist who wrote under the pseudonym "Hibernicus." Wood Sorrel *(Oxalis montana)* is represented by carpets of shamrocklike foliage and pretty pink-striped white blossoms in late spring and summer. More modest than any of these is one of our native orchids: the rare Lesser Rattlesnake Plantain (*Goodyera repens* var. *ophioides*), which is smaller than the common *G. pubescens* of lower elevations and has a one-sided raceme of flowers. It is found at scattered stations throughout the high mountains.

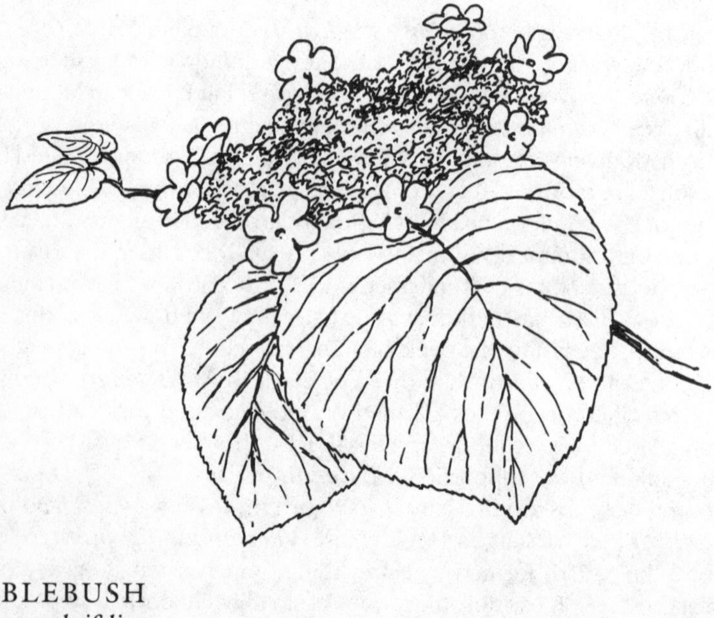

HOBBLEBUSH
Viburnum alnifolium

BLUEBEAD LILY
Clintonia borealis

One wildflower that grows nowhere in the world except in the spruce-fir forests of the Great Smoky Mountains is Rugel's Indian Plantain (*Cacalia rugelia,* formerly named *Senecio rugelia*). It is a colonial plant, sometimes covering extensive areas with its large leaves, and blooms in late summer. The yellow and brown inflorescence consists of tubular florets with protruding stigmas and is completely devoid of the petallike rays that are found in so many other composites.

The forager in these woods will be delighted to learn that the Blackberries *(Rubus canadensis),* which can be expected wherever there is a path or other opening, are thornless at these high elevations.

With 90 inches or more of precipitation annually, and with deep shade and cool temperatures retarding evaporation, this is an extremely moist environment. The damp black humus beneath a layer of fallen needles is springy underfoot. Rocks and downed trees are overlaid with sheets of mosses. Ferns thrive here too, and the trees are adorned with many kinds of lichens.

In these surroundings sounds tend to be absorbed, and one becomes acutely aware of small creatures nearby, like the black-capped chickadee that replaces the Carolina chickadee of lower altitudes and the juncos that have moved high up into the mountains to nest. One needs no such quiet, though, to notice two of the forest's best known and most vociferous inhabitants—the red squirrel and the common raven.

CLINGMAN'S DOME

Although the southern Canadian zone forests are situated atop mountain peaks, many of them are easily accessible during most of the year. A good example is Clingman's Dome, on the North Carolina–Tennessee state line, which at 6,643 feet is the highest point in Great Smoky Mountains National Park. A paved spur road from Newfound Gap at the Tennessee–North Carolina line on U.S. 441 takes you to a parking lot from which there is a half-mile trail through aromatic firs and spruces to an elevated observation deck on the summit. This structure has been decried by some as having degraded the landscape and defended by others as providing the only means of viewing the surrounding mountains. The latter have a point: Because of the latitude the southern mountains have no timberline and are forested right to the top.

Among the herbaceous flowers to be found in these woods are Painted Trillium in early spring, and the robust Filmy Angelica *(Angelica triquinata)* which in August attracts myriads of bees to its flat clusters of greenish flowers. Other late bloomers are *Hypericum graveolens,* one of the most beautiful of the St. John's-worts, and White Snakeroot *(Eupatorium rugosum).* Far from being confined to the spruce-fir community, White Snakeroot is one of the most numerous of all the wildflowers in the southern mountains. When you consider the multitudes of asters, goldenrods, and other composites in the race for that distinction, this is saying quite a lot.

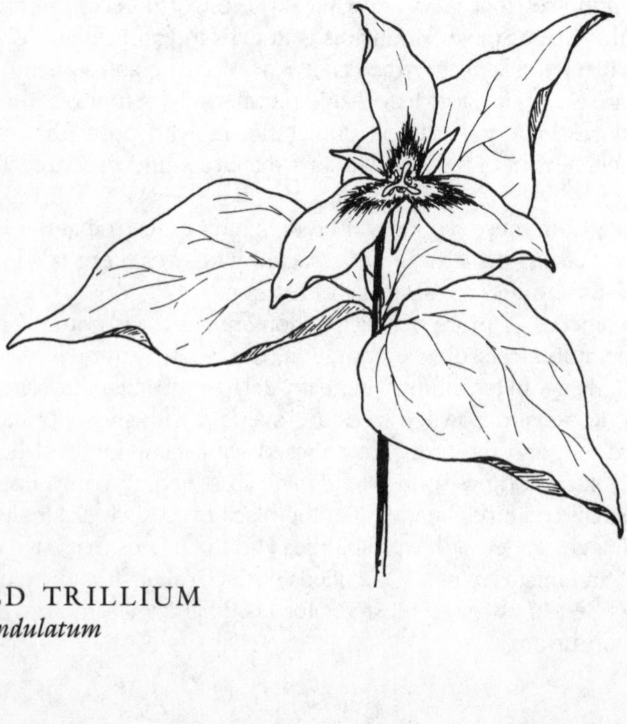

PAINTED TRILLIUM
Trillium undulatum

At the shrub level, Hobblebush is joined by Red-berried Elder *(Sambucus pubens).* This relative of the edible Black Elderberry *(S. canadensis)* has similar foliage, but the inflorescence is pyramidal and the fruits are scarlet. It is not likely, therefore, that the two species will be confounded, which probably is a good thing since *S. pubens,* if not actually toxic, has been suspected of causing ill effects.

One of the most striking sights comes in autumn when the scarlet fruits of American Mountain Ash stand out in vivid contrast to the background of dark conifers.

A short distance below Clingman's Dome, on the eastern side of the spur road, is the Spruce-Fir Self-guiding Nature Trail. This is a short circular walk, and a good place to enjoy Bluebead Lily and Hobblebush, both of which grow here in profusion.

Directly across the road one can sample the Appalachian Trail in either direction. In addition to Wood Sorrel and Rugel's Indian Plantain, the red form of Wake Robin *(Trillium erectum)* will be found here. It is interesting that in the southern mountains most plants of this species, particularly at lower elevations, bear white flowers; confusion with other Trilliums can be avoïded, however, by remembering that *T. erectum* has a maroon ovary irrespective of the petal color.

For moss enthusiasts this section of the Appalachian Trail has several species, the most attractive of which is probably the Stairstep Moss *(Hylocomium splendens).* This is one of several mosses that resemble ferns in miniature and owes its common name to the fact that each year's "fronds" arise from the middle of the preceding year's growth. Lichens are everywhere, and the light gray, threadlike festoons of Old Man's Beard (*Usnea* spp.) hang from the trees much in the manner of Spanish Moss but on a smaller scale.

ALONG THE PARKWAY

The loftiest point on the Blue Ridge Parkway is reached near mile 431. Here at the Haywood Jackson Overlook there is a nature trail leading over the top of Richland Balsam, the 6,410-foot mountain that is the apex of the Great Balsam range. The trail is a 1.4-mile loop with a 400-foot gain in elevation, not strenuous except possibly for some not accustomed to the altitude.

In just a few years Richland Balsam has changed from being the quintessential Fraser Fir forest to being a place to view the apparently inexorable destruction of these trees by the balsam woolly aphid, a European pest against which no defense has yet been found, and to observe the rapidity with which vegetation change takes place in the wake of such a calamity.

Thornless Blackberries, ever the opportunists, are striving vigorously to fill the void in company with such other sun seekers as Red-berried Elder, Mountain Ash, Filmy Angelica, and Whorled Wood Aster *(Aster acuminatus).* The last is the earliest of all the asters to bloom, and although its leaves do appear to be

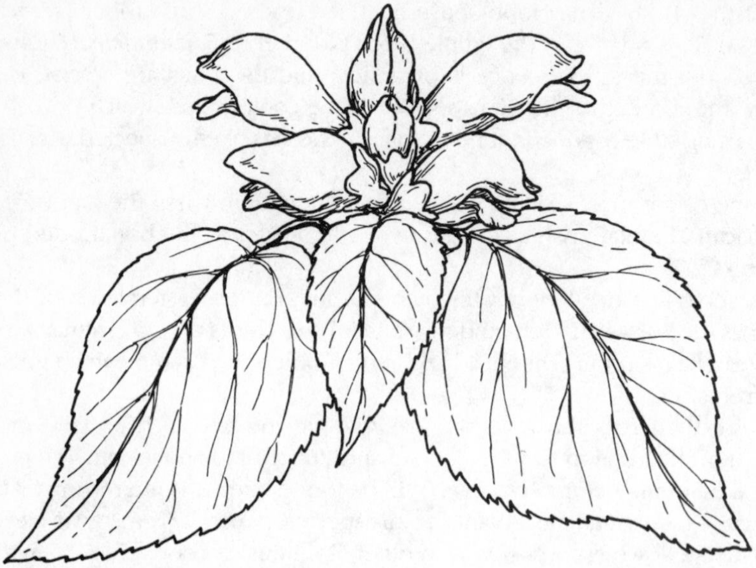

PINK TURTLEHEAD
Chelone lyoni

disposed in whorls as the name would suggest, in actuality they are arranged alternately (a fine point, but one that could save you from heading down a blind alley in a wildflower guide book). The red stems and numerous racemes of tiny white flowers of Fringed Bindweed *(Polygonum cilinode)* call attention to that rather unprepossessing plant as they scramble over whatever support they can find, while isolated patches of Bluebead Lily and Pink Turtlehead *(Chelone lyoni)* consolidate their holdings along the trail edges.

Southern Lady Fern *(Athyrium asplenioides)* and Haircap Moss *(Polytrichum commune)* are plentiful, and even the brilliant orange Witches' Butter fungus (*Dacrymyces* sp.) finds a niche here—on the boles of the dying firs.

Nearer to the Smokies, at mile 451.2, is Waterrock Knob. Although it is slightly lower than Richland Balsam, it affords superior views, which are especially enhanced by the Mountain Ash trees in September, when they are in brilliant fruit. At one point the path to the top skirts a grassy area with an abundance of Purple Fringed Orchid *(Platanthera psycodes)* in mid-June. Much less glamorous, but with a comeliness and fragrance all its own, is the orange-stemmed, white-blossomed Dodder (*Cuscuta* sp.), which twines about other vegetation all along the sunlit trail. Dodder begins life like any other plant whose seed germinates in the soil, but as soon as it finds a suitable host it attaches itself to the stem, extracting its food supply by means of sucking organs called haustoria. Once this parasitic relationship is established, the lower portion of the Dodder's stem atrophies, severing its connection with the ground.

Another example of a boreal forest with superb viewpoints is Devil's Courthouse. This impressive rock profile can be best seen from the parking area at mile 422.4, and is reached by an easy trail through an open growth of spruce and fir. The exposed crest (elevation 5,720 feet) supports a number of interesting shrubby plants, among them Sand Myrtle *(Leiophyllum buxifolium),* Maleberry *(Lyonia ligustrina),* Black Chokeberry *(Sorbus melanocarpa),* Dwarf Prairie Willow (*Salix humilis* var. *microphylla*), and Wine-leaved Cinquefoil *(Potentilla tridentata).* The last named is definitely a northern plant, and although it is common at sea level in maritime Canada, it extends into the southern Appalachians only at scattered locations on the higher mountains. It has the distinction of being our only *Potentilla* with white flowers instead of yellow. The specific name refers to the three teeth at the end of each leaflet. Each leaf has three of these leaflets (although "Cinquefoil" would suggest five). The term "Wine-leaved" is appropriate, though, since the leaves do turn rich red in fall.

From the southern end of the adjacent parking turnout (at mile 423.6, identified as Courthouse Valley Overlook), one may explore the wooded slopes of Tanasee Bald, which is the southernmost point reached by the Blue Ridge Parkway. In spring the grassy trailsides are brightened by drifts of Trout Lily *(Erythronium americanum)* and Wood Anemone *(Anemone quinquefolia).* The Trout Lily is much more to be expected in low-lying wet areas, but the little brook here apparently provides a satisfactory habitat.

In summer there is a species of Wild Bergamot *(Monarda clinopodia),* a coarse plant of the Mint family with dense heads of long, narrow flowers that may be white or pale pink, finely dotted with purple. Several *Monardas* are native to the region, but this is the one most prevalent in the cooler mountains.

Tassel-rue *(Trautvetteria carolinensis)* is an attractive wildflower that the South shares with several of the prairie states. It is tall, with large, deeply divided leaves and clusters of small flowers composed of numerous white stamens surrounding a central group of green pistils (there are no petals or sepals at anthesis). Near the opposite end of the scale is a little plant with a big name: Dwarf Enchanter's Nightshade *(Circaea alpina).* Less than a foot high and with only a few insignificant white flowers, it can easily be overlooked, but it is worth seeking out, if only as a curiosity. Dwarf Enchanter's Nightshade and other members of its genus are unique in that all of their floral parts are in sets of two.

American Beech trees *(Fagus grandifolia)* are an important component of the woods on Tanasee, and furnish abundant food for the wildlife of the mountain community.

MOUNT MITCHELL

As the highest peak east of the Mississippi River, Mount Mitchell (6,684 feet) would be expected to contain an exemplary southern coniferous forest, and it certainly does not disappoint. The summit is virtually pure Fraser Fir, and although depredation by the balsam woolly aphid is evident, the young trees apparently are less susceptible to attack by this insect and it is hoped that some of them may possibly provide resistant genetic material. In addition to the firs, though, several other kinds of trees have also been dying on Mount Mitchell (and in many other places as well), and it is suspected that air pollution is part of the problem. At any rate, intensive studies are under way to determine what factors are responsible.

Mount Mitchell is a state park, and the summit can be reached via N.C. 128, which intersects the Blue Ridge Parkway at mile 355.3. In the open area surrounding the observation tower there can be seen *Hypericum mitchellianum,* the St. John's-wort that commemorates Dr. Elisha Mitchell, who lost his life in 1857 while exploring the mountain that also bears his name.

A nature trail threads through dark fir woods brightened by a few species of plants which, as though to compensate for lack of diversity, bloom in profusion. From rocky seeps arise clouds of Michaux's Saxifrage *(Saxifraga michauxii).* If ever there was a flower that needed to be seen through a hand lens in order to be fully appreciated, this is it. Of the five small white petals, the upper three are of a different shape than the lower two (which alone makes it distinctive among our Saxifrages) and are marked with greenish yellow glands; set off against these are ten stamens with knoblike anthers of vivid red-orange. Wood Sorrel is an abundant ground cover, and where the canopy is interrupted masses of White Snakeroot take advantage of the sunlight.

August is a delightful time to explore the trail leading from the picnic grounds to Mount Craig. At the beginning this takes you through natural nurseries of bright green young firs intermixed with Pink Turtlehead and Whorled Wood Aster, while out on the open ridge the large-flowered Goldenrod *(Solidago glomerata)* is dazzling against the backdrop of hazy blue mountain ranges. The delicious Mountain Red Raspberry (*Rubus idaeus* var. *canadensis*) has a narrow distribution, but this is one place where it flourishes.

OTHER PLACES TO EXPLORE

There are many other examples of southern spruce-fir forests in more remote regions of eastern Tennessee and western North Carolina, where they are beyond the range of the casual tourist but can be enjoyed by the seasoned hiker or backpacker.

Among those that should not be overlooked are Shining Rock Wilderness, with its added attraction of gleaming white quartz outcrops, and numerous high peaks in the Great Smokies (there are 16 higher than 6,000 feet). Among the latter Mount Le Conte, on the Tennessee side of the National Park, is the undisputed favorite. Myrtle Point, a rocky promontory at the top clothed in and named for Sand Myrtle, is just one of its outstanding features.

The botanically rich Grandfather Mountain was a favorite stamping ground of Harvard's Asa Gray. One of his early predecessors, André Michaux, in a burst of Gallic enthusiasm, proclaimed it to be the highest peak in North America; it isn't even close, of course, but at 5,938 feet it *is* the climax of the Blue Ridge escarpment. It is now privately owned, but the public may use its trails by permit; access is from U.S. 221 near Linville, N.C.

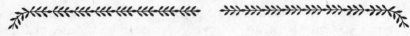

7

Northern Hardwoods in the South

THE CANADIAN zone forests have been characterized as bands of distinctive vegetation wrapped around the upper reaches of the Southeast's highest mountains, but it must not be inferred that these belts are so regular that their lower edges follow a constant elevation like contour lines on a topographic map. On the contrary, complex environmental factors cause considerable fluctuations, so that in some places segments of fir forests extend quite far down the slopes while elsewhere concentrations of deciduous trees migrate upward into seemingly unsuitable territory.

Below this jagged boundary is a narrow intermediate, or transitional, zone. Firs are absent here, and although Red Spruce *(Picea rubens)* is present in quantity and Carolina Hemlock *(Tsuga caroliniana)* in much smaller numbers, these forests are basically deciduous and have been given the name "northern hardwoods" irrespective of where they occur.

Many of the broad-leaved trees commonly associated with the spruce-fir complex—Yellow Birch *(Betula lutea),* American Beech *(Fagus grandifolia),* Mountain Maple *(Acer spicatum),* and American Mountain Ash *(Sorbus americana)*—are still with us, but they are joined by such others as Wild Black Cherry *(Prunus serotina),* Northern Red Oak *(Quercus rubra),* Yellow Buckeye *(Aesculus octandra),* Sugar Maple *(Acer saccharum),* Hawthorn (*Crataegus* spp.), Mountain Holly (*Ilex ambigua* var. *montana*), and Serviceberry *(Amelanchier arborea).* Noticeably absent are such indicators of cove forests as Tulip Tree and Silverbell. Shrubs include Minnie-bush *(Menziesia pilosa),* Mountain Pepperbush *(Clethra acuminata),* and Hazel (*Corylus* spp.).

Few species of herbaceous wildflowers can tolerate the dense, cool shade of the Canadian zone forests, but in these predominantly deciduous woods with their greater quota of sunlight there is much more variety.

Since the northern hardwoods share the higher elevations with the spruce-fir community, western North Carolina and eastern Tennessee are replete with

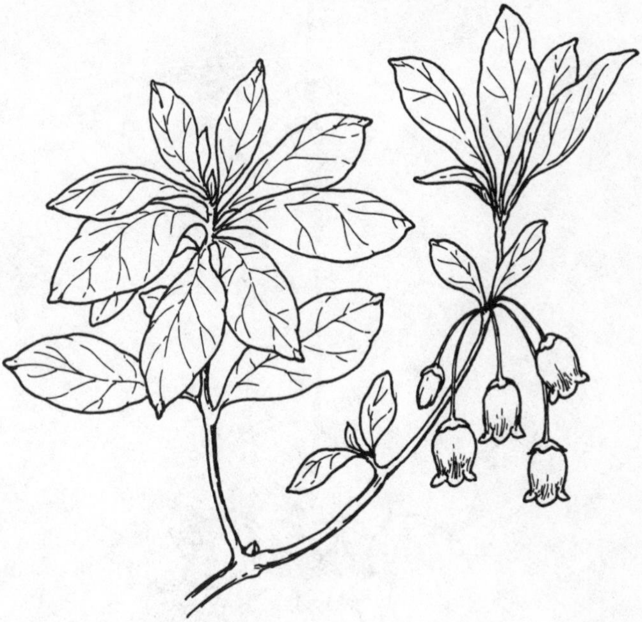

MINNIE-BUSH
Menziesia pilosa

examples, and again the Blue Ridge Parkway and Great Smoky Mountains National Park serve us well in affording access to heretofore remote regions.

MOUNT PISGAH

Mount Pisgah (5,721 feet) was the centerpiece of George W. Vanderbilt's 100,000-acre estate, and is now a prominent feature of Pisgah National Forest. The trail to the summit starts from the north end of a parking lot off the Blue Ridge Parkway at mile 407.7. The round trip is 2.4 miles, with easy going at first, but then the trail becomes progressively steeper.

Not far from the beginning there are vines of Dutchman's Pipe *(Aristolochia macrophylla)* climbing trees that are rooted below trail level on the left-hand bank. This perspective affords an especially good opportunity to study the unusual flowers, which all too often are far above one's head. Nearby are a number of Mountain Pepperbush shrubs, which are conspicuous even when out of flower by their flaky cinnamon-colored bark.

A little farther along you can expect to find Hobblebush *(Viburnum alnifolium)* and Southern Mountain Cranberry *(Vaccinium erythrocarpon),* as well as an abundance of Filmy Angelica *(Angelica triquinata).* Some of the taller herbs

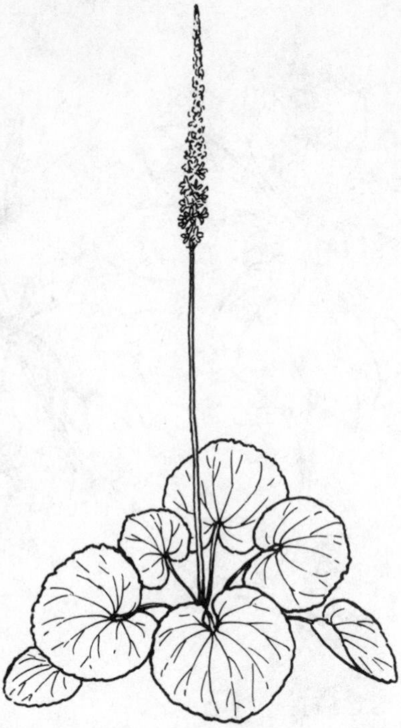

GALAX
Galax aphylla

are Featherbells *(Stenanthium graminifolium),* Turk's-cap Lily *(Lilium superbum),* and Black Cohosh *(Cimicifuga racemosa).* As you approach the top, Galax *(Galax aphylla)* and Wine-leaved Cinquefoil *(Potentilla tridentata)* are very much in evidence.

The summit, which has a raised deck for viewing the mountain panorama, has a dense shrub cover, including Blueberries (*Vaccinium* spp.), Black Huckleberries *(Gaylussacia baccata),* Southern Bush Honeysuckle *(Diervilla sessilifolia),* and Beaked Hazel *(Corylus cornuta).* Another interesting shrub occurring here is the Dwarf Prairie Willow (*Salix humilis* var. *microphylla*).

The plenitude of berries together with the tantalizing presence of other food at the nearby campground attracts black bears to this area, and signs of this might be noticed along the trail, but the animals themselves are rarely sighted.

SHUT-IN TRAIL

To reach his rustic hunting lodge at Buck Spring, near the present Pisgah Inn, Vanderbilt and his guests rode on horseback the 17 miles from Biltmore House, his baronial estate in Asheville, N.C. The route they followed now passes through both national park and national forest lands, and it has been reconstructed by the two agencies as the Shut-in Trail, so called because of the overhead canopy formed by the trees. Much of it parallels the Blue Ridge Parkway between miles 405.5 and 393.7, connecting overlooks that are spaced from 1 to 3 miles apart. It is possible to enter the trail at some of these overlooks and cover one or more segments according to the time one has available.

Botanically speaking, one of the most rewarding sections is the one starting at Mills Valley Overlook (mile 404.5) and heading north. After about a 10-minute walk you suddenly come out into an open glade filled in early August with hundreds of Turk's-cap Lilies *(Lilium superbum),* many of them towering overhead, swarming with tiger and spicebush swallowtails and other butterflies.

Offsetting the heavy-looking pendent bells of the lilies are soft, misty domes of mauve Joe-Pye-Weed *(Eupatorium fistulosum)* and the upright frosty white spires of Black Cohosh *(Cimicifuga racemosa),* while the bases of all these tall plants are banked with golden yellow Ox-eye *(Heliopsis helianthoides).* Color accents are provided by blue-violet Spiderwort *(Tradescantia subaspera)* and deep pink *Phlox carolina,* and unusual flower forms by Leatherflower *(Clematis viorna),* Pale Jewel Weed *(Impatiens pallida),* and Wild Bergamot *(Monarda clinopodia).*

There are no other displays to compare with this either along the trail to the north or in the sections of it to the south of this one, but a number of other interesting species will make further exploration worthwhile. There are, for instance, small populations of Fringed Campion *(Silene ovata),* Bunchflower *(Melanthium hybridum),* Hairy Angelica *(Angelica venenosa),* and Lopseed *(Phryma leptostachya),* as well as some scattered Carolina Lilies *(Lilium michauxii).* The Small-flowered False Hellebore *(Veratrum parviflorum)* is plentiful, although not so attractive as *V. viride.* Among some of the other wildflowers of modest appearance to be found here in late summer are Rough Avens *(Geum virginianum),* Maryland Figwort *(Scrophularia marilandica),* Enchanter's Nightshade *(Circaea quadrisulcata),* Agrimony *(Agrimonia gyrosepala),* Indian Tobacco *(Lobelia inflata),* and Flowering Spurge *(Euphorbia corollata).*

One always wonders whether some unusual springtime feature may have preceded such a spectacular summer pageant. In this case the forest floor in May is covered with a luxuriant mass of Interrupted Fern *(Osmunda claytoniana)* that easily steals the show from the usual proliferation of early wildflowers.

BIG BUTT TRAIL

In contrast to the popularity of Mount Pisgah, this area is relatively unknown, but the Big Butt Trail (or at least the first few miles of it) deserves to be seen in the spring by everyone who is interested in the flora of the southern mountains. It begins just south of the Balsam Gap parking area at mile 359.9 on the Blue Ridge Parkway; be sure to take the narrow foot trail, not the old logging roadbed.

Immediately you come upon masses of Wake Robin *(Trillium erectum),* Spring Beauty *(Claytonia caroliniana),* and False Solomon's-seal *(Smilacina racemosa).* There are dozens of other wildflowers that are more modest in number but still plentiful, including many that have a preference for cool, high elevations, such as Canada Mayflower, Painted Trillium, Bluebead Lily, Dwarf Enchanter's Nightshade *(Circaea alpina),* Pink Turtlehead *(Chelone lyoni),* Squirrel-corn *(Dicentra canadensis),* and Rose Twisted-stalk *(Streptopus roseus).*

Wild Leek, or Ramp *(Allium tricoccum),* is abundant. Where it has managed to escape being dug in early May for its edible bulbs, it will put up a flower stalk in summer, long after its broad leaves have vanished. For those who have a special appreciation of the so-called lower plants, the Big Butt Trail also has a lush growth of mosses, club mosses, and ferns.

CRAGGY GARDENS

Craggy Gardens is a large heath bald that extends over several eminences bearing such names as Craggy Knob, Craggy Pinnacle, and Craggy Dome. Its compact clumps of Catawba Rhododendron separated by neat pathways suggest "gardens." The bald itself will be discussed in Chapter 9, which is devoted to the

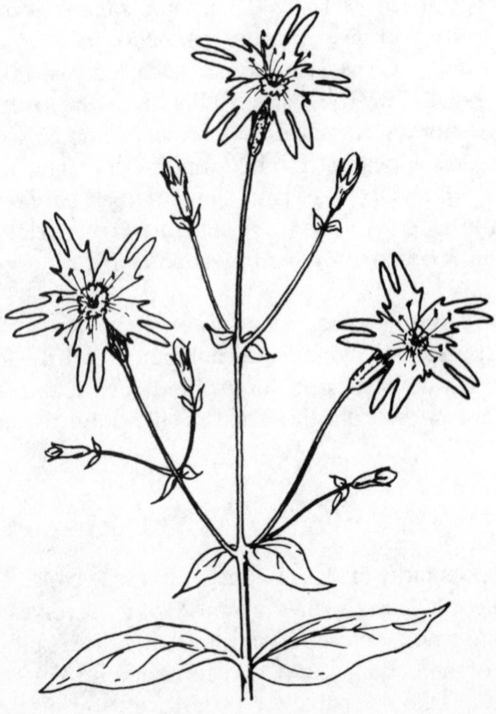

FIRE PINK
Silene virginica

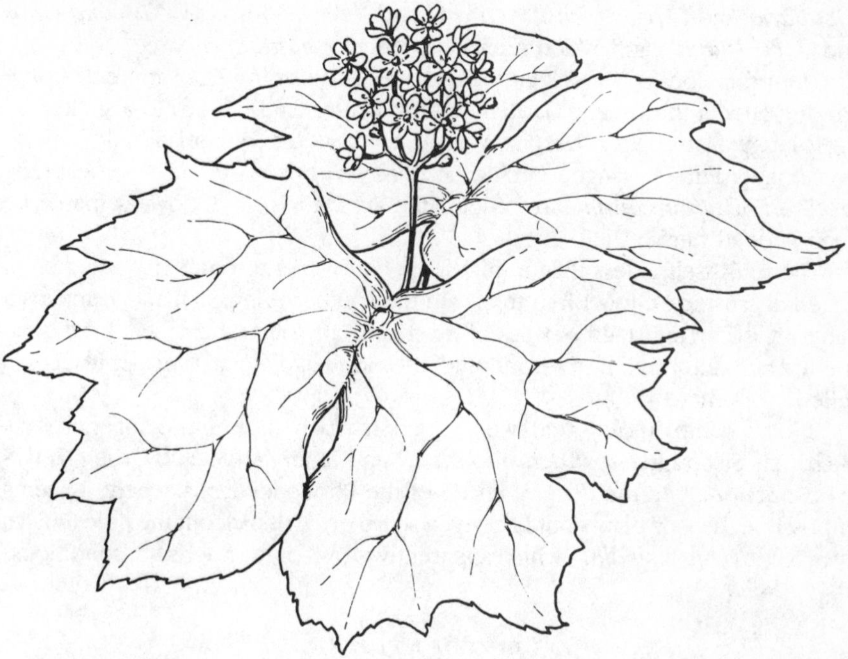

UMBRELLA LEAF
Diphylleia cymosa

phenomenon of mountain balds. What is of interest to us here is the northern hardwoods community that lies just below the balds in two areas in particular.

One is known as the Craggy Gardens Recreation Area. It is reached by a short spur road from the Blue Ridge Parkway at mile 367.7, which is itself a wildflower garden. The first part of the road has wide sunny banks on each side, and these are covered at various times of the year with mixtures of Fire Pink *(Silene virginica),* Tickseed *(Coreopsis pubescens),* Obedient Plant *(Dracocephalum virginianum),* Flowering Spurge, Black-eyed Susan *(Rudbeckia hirta),* Robin's Plantain *(Erigeron pulchellus),* Columbine *(Aquilegia canadensis),* and many others. Beyond this, where it is somewhat shaded by trees, there is Umbrella Leaf *(Diphylleia cymosa),* Pale Jewel Weed, and Spiderwort. The road ends at a large parking area adjacent to a picnic area with groves of Yellow Buckeye mixed with Hawthorns and Mountain Ash. Roan Rattlesnake Root *(Prenanthes roanensis),* a relatively rare member of the genus, can be found here.

If possible, you should try to time a visit for late April or early May. After parking at the picnic grounds, walk a short distance down the entrance road and explore the woods on the left-hand side. This slope is liberally carpeted with Wake Robin, Spring Beauty, Dutchman's Breeches and Squirrel-corn (*Dicentra cucullaria* and *D. canadensis*), Canada and Yellow Violets (*Viola canadensis* and *V. eriocarpa*), and Wood Anemone *(Anemone quinquefolia).*

The other locale is a self-guiding nature trail leading from the end of the parking area at the Craggy Gardens Visitor Center (mile 364.6) to a shelter on the Craggy Flats Bald. The grassy road bank at the bottom should not be overlooked; Purple Fringed Orchid *(Platanthera psycodes)* and Waterleaf *(Hydrophyllum virginianum)* are among the uncommon wildflowers that have been spotted here.

The trail itself is less than a mile long, and winds through an "orchard" of lichen-encrusted Yellow Birch trees. (In the southern Appalachian Mountains a naturally occurring stand of trees of practically any one species that have been stunted and deformed by the wind until they look like old gnarled apple trees is called an orchard.)

It is a damp, mossy trail well suited to such moisture-loving herbs as Michaux's Saxifrage *(Saxifraga michauxii)* and Mountain Meadow-rue *(Thalictrum clavatum).* In the deepest shade of the Rhododendrons, where it seems impossible that any plant could flower, a sharp eye will pick out the little Small's Twayblade *(Listera smallii),* which apparently grows only under such conditions.

NEWFOUND GAP

Northern hardwoods forests occur in much of the territory embraced by the Great Smoky Mountains National Park, but there can hardly be a more convenient place to sample them than at Newfound Gap. This is the point near the center of the park at which the Appalachian Trail crosses U.S. 441 on its way from Maine to Georgia.

The altitude at the intersection is 5,040 feet above sea level, and a short walk along the trail either north or south will disclose flora typical of the hardwoods zone (but hike several miles in either direction and you will find that you have ascended into the spruces and firs of the Canadian zone).

In the spring there are spectacularly large patches of Fringed Phacelia *(Phacelia fimbriata)* and Spring Beauty. Dutchman's Breeches, Wake Robin, and Northern White Violets *(Viola pallens)* also are abundant, and cushions of tiny Thyme-leaved Bluets adorn the edges of the trail. Later in the year, along with the crimson fruits of Red-berried Elder *(Sambucus pubens),* comes that most attractive of the native St. John's-worts *(Hypericum graveolens),* followed in the fall by extensive beds of White Snakeroot *(Eupatorium rugosum).*

8

Cove Forests and Their Wildflowers

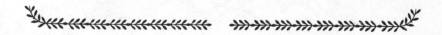

USING THE Great Smokies as an example, we have seen how numerous elements have combined to make the southern Appalachians an ideal home for an extraordinary variety of plant species. Within this favored environment, as we move down to altitudes under 4,500 feet, and in some places well below 3,000 feet, we come into what is, botanically speaking, the richest part of the entire eastern deciduous forest. Here there is a unique feature that stands out above all others and, fortunately, is found throughout the southern mountains. It is known as the cove hardwood forest. The word "cove" can mean a wide, flat valley rimmed by mountains (Cades Cove in the Great Smoky Mountains National Park is an example of this), but in the present context it is a place where, over a very long time, a stream has cut a V-shaped groove deeper and deeper into a mountainside, creating a sheltered recess with gently sloping sides and depositing a burden of rich organic material near the bottom.

If a single word could be used to characterize the flora of a cove forest it would have to be "diversity," and this is perhaps best illustrated by its trees. It is possible to find 40 or more species in a single cove, none of which is dominant.

To many of the so-called northern hardwoods are added Tulip Trees *(Liriodendron tulipifera),* a close relative of the magnolias but often mistakenly called Yellow Poplar; White, Black, and Chestnut Oaks *(Quercus alba, Q. velutina,* and *Q. prinus);* Hickories (*Carya* spp.); Black Walnut *(Juglans nigra);* Basswood *(Tilia heterophylla);* White Ash *(Fraxinus americana);* Black Cherry *(Prunus serotina);* Mulberry *(Morus rubra);* and Black Locust *(Robinia pseudoacacia).* Trees with showy blossoms are a hallmark of the southern Appalachians, and here we have no fewer than three deciduous Magnolias, also Silverbell *(Halesia carolina),* Sourwood *(Oxydendrum arboreum),* and in the understory Flowering Dogwood *(Cornus florida).*

TULIP TREE
Liriodendron tulipifera

The light shade and moisture of the cove forests are ideal for Rosebay Rhododendron *(Rhododendron maximum),* and it is beyond doubt the commonest evergreen shrub. In deep, wet ravines Eastern Hemlock *(Tsuga canadensis)* and Dog Hobble (*Leucothoe axillaris* var. *editorum*) are prolific, while the drier sections favor Mountain Laurel *(Kalmia latifolia)* and Wild Hydrangea *(Hydrangea arborescens).*

Impressive though this cataloging of woody plants is, these fertile, well-watered coves are also responsible for the southern Appalachians' having become deservedly famous as one of the richest natural wildflower gardens in the world outside of the tropics. To attempt a recitation of species here would only invite repetition (the list would run to well over a thousand); it will be more meaningful to attribute significant ones to specific areas as they are singled out for exploration.

IN THE SMOKIES

There could be no better place to see cove hardwoods than Great Smoky Mountains National Park, where there are many examples at hand.

Greenbrier Cove is an accessible but relatively undisturbed section. The entrance is off Tenn. 73 about 6 miles east of Gatlinburg, and after 3 miles on this road (Greenbrier Road) you come to a fork. The left prong takes you to Ramsay Cascades—the park's highest waterfall—on a rather strenuous but rewarding hike. In the first mile there are large specimens of Silverbell, Tulip, and Cucumber *(Magnolia acuminata)* trees. On the right is the Porters Creek Trail, which might well be named the Jekyll-Hyde Trail. During the first 4 miles it follows the stream through a delightful mature hardwoods forest, climbing only gradually, but after that it becomes the most difficult and dangerous trail in the park.

The entrance road and the initial segments of both trails are recommended for their profusion of wildflowers in the spring. Several species of Trillium and numerous Violets plus Toothwort *(Dentaria diphylla),* Foamflower *(Tiarella cordifolia),* White Fringed Phacelia *(Phacelia fimbriata),* and Crested Dwarf Iris *(Iris cristata)* brighten the woods. On the banks Showy Orchis *(Galearis spectabilis),* Stonecrop *(Sedum ternatum),* and a long-sepaled form of Wild Ginger *(Asarum canadense* var. *acuminata)* are common. Lettuce Saxifrage *(Saxifraga micranthidifolia)* stands in the middle of rushing brooks, while the aromatic maroon flowers of Sweet Shrub *(Calycanthus floridus)* are seen along the streamsides.

One of the richest concentrations of cove vegetation is located on U.S. 441 at the Chimneys picnic area, where you can take the self-guiding Cove Hardwood Nature Trail through second-growth and virgin forest for three-quarters of a mile. (If you are coming from the direction of Gatlinburg, look for Yellowwood trees *(Cladrastis lutea)* and Umbrella Leaf *(Diphylleia cymosa)* along the road as you approach the site.) Many of the spring flowers noted in Greenbrier Cove are here, with Yellow Trillium *(Trillium viride* var. *luteum)* plentiful near the bottom and the white form of *T. erectum* forming large drifts higher up. The vivid color of massed Creeping Phlox *(Phlox stolonifera)* contrasts with the tiny

CRESTED DWARF IRIS
Iris cristata

yellow flowers of Clustered Snakeroot *(Sanicula gregaria)* and the equally small white blossoms of Sweet Cicely *(Osmorhiza claytoni).*

Level terrain adds special appeal to the Kephart Prong Trail. This takes off from the east side of Newfound Gap Road (U.S. 411) about 7 miles north of the Oconaluftee Visitor Center, and crosses a bridge immediately after leaving the highway.

In an open area near the beginning there is a population of a small-flowered Avens, *Geum vernum,* which is of interest as being one of those plants that has migrated from the American Midwest. As you continue along, watch for the path to turn sharply to the left across a small log footbridge. On the other side, at a patch of the leafy Cream Violet *(Viola striata),* the trail goes right on an old roadbed. But first look to the left, where the open woods in late April are white as snow with Fringed Phacelia *(Phacelia fimbriata).* Several spring wildflower species that habitually form extensive drifts of blooms grow along Kephart Prong; these include Creeping Phlox *(Phlox stolonifera),* Wood Anemone *(Anemone quinquefolia),* Canada Mayflower *(Maianthemum canadense),* Spring Beauty *(Claytonia caroliniana),* and Corn Salad (*Valerianella* spp.), but Fringed Phacelia is the most spectacular.

Along wet stretches of the road there are Jack-in-the-Pulpit *(Arisaema triphyllum)* and Sweet White and Marsh Blue Violets *(Viola blanda* and *V. cucullata).* Another three plants with tiny flowers—Pennsylvania Bitter Cress *(Cardamine pensylvanica),* the creeping Golden Saxifrage *(Chrysosplenium americanum),* and the tall Lettuce Saxifrage—actually get their feet wet in small streams and seeps.

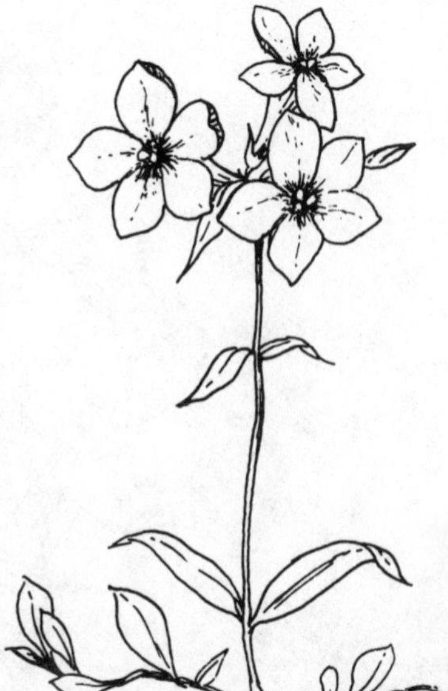

CREEPING PHLOX
Phlox stolonifera

There are, as you would expect, dozens of upland wildflowers, such as White Baneberry *(Actaea pachypoda),* Meadow Parsnip *(Thaspium trifoliatum)* in both yellow-flowered and purple-flowered forms, Bellwort *(Uvularia sessilifolia),* Mountain Chickweed *(Stellaria pubera),* Wild Ginger *(Asarum canadense),* Showy Orchis, and several Trilliums. In fact, in a place like this where one can see more than 60 species in bloom in a single day, it is sometimes hard to think of a spring wildflower that is *not* here.

PISGAH NATIONAL FOREST

With 4,000 square miles of southern mountain land and countless streams rushing down its slopes, Pisgah National Forest has cove hardwoods just about everywhere. This makes it easy to choose a trail that is accessible, short, and satisfying, and has a pleasant surprise at its destination. Such a trail is the one to Moore Cove.

The trailhead is on N.C. 276 just 8.2 miles south of the Blue Ridge Parkway, next to a bridge over Looking Glass Creek and beneath a battered Fraser Magnolia. After passing through ranks of Shrub Yellowroot *(Xanthorhiza simplicissima),* the trail climbs briefly to an old logging road, and this is where the wildflower show begins.

There are many Violets, the season starting with the Yellow Round-leaved and Halberd-leaved species *(Viola rotundifolia* and *V. hastata).* These are followed by Canada *(V. canadensis);* Common Blue, including the grayish form, *V. papilionacea* var. *priceana,* also known as Confederate Violet; Three-lobed *(V. triloba);* and still another yellow one, *V. eriocarpa.* Trilliums include Wake Robin *(T. erectum)* and Nodding *(T. cernuum).*

On the road banks are the large-flowered Heartleaf *(Hexastylis shuttleworthii),* the parasitic Cancer-root *(Orobanche uniflora),* Great Yellow Sorrel *(Oxalis grandis)* with purple-edged leaflets, Foamflower *(Tiarella cordifolia),* and the semiwoody Heart's-a-bustin' *(Euonymus americanus).* Ginseng *(Panax quinquefolium)* is seen occasionally, but people do not seem to be able to leave it alone and it seldom survives for very long near trails as popular as this one.

Orchids are well represented, with Rattlesnake Plantain *(Goodyera pubescens)* and Crane-fly Orchid *(Tipularia discolor)* both plentiful. A special treat is in store for those who happen to be here on one of the rare August mornings when the dainty Three Birds Orchid *(Triphora trianthophora)* puts out its ephemeral flowers, for it too is abundant along this trail.

As you come over a rise about three-quarters of a mile from the start, you are greeted with the sight of a lovely waterfall dropping from a high projecting ledge. If you wish, you can walk behind the falling stream of water to inspect the ferns, mosses, and liverworts that grow in the constant mist.

This 14-mile section of N.C. 276, extending south from Wagon Road Gap on the Blue Ridge Parkway through Pisgah National Forest toward Brevard, is extremely popular. Much more than just a beautiful woodland drive, it includes a stretch of the Davidson River—one of the finest trout streams in the East—Looking Glass Falls, Sliding Rock, and the Cradle of Forestry. Just driving along the road one can glimpse many showy wildflowers, such as Fire Pink *(Silene virginica),* Sundrops *(Oenothera tetragona),* Yellow Fringed Orchid *(Platanthera ciliaris),* Spiderwort *(Tradescantia subaspera),* and Goatsbeard *(Aruncus dioicus),* and superb displays of Rosebay Rhododendron.

JOYCE KILMER–SLICKROCK WILDERNESS

It is generally accepted that the true American wilderness as it was known to the early white settlers no longer exists anywhere, but because we want to preserve those fragments of our land that most resemble what has been lost, we keep redefining our notion of wilderness to fit what we have left. So it is that we have—and are grateful for—relatively pristine forests such as those in Joyce Kilmer–Slickrock Wilderness. This tract of nearly 15,000 acres is in North Carolina's Nantahala National Forest, 12 miles west of Robbinsville.

A small but significant part of the area was set aside in 1936 as the Joyce Kilmer Memorial Forest, a memorial to the poet who wrote "Trees." Here a trail consisting of two loops joined to form a figure eight leads one through a magnificent virgin forest with immense Hemlocks, Tulip Trees, Sycamores, Basswoods, Beeches, and Cherries, some of them centuries old, 20 feet around at the base, and a hundred feet high. Very high rainfall feeds Little Santeetlah Creek, which flows through the area, and contributes to a "rain forest" environment that is especially conducive to the growth of mosses, lichens, and ferns.

The trees and shrubs are so dense and the amplitude of their foliage so great that spring is the only time that is propitious for wildflowers. Among the earliest to emerge are Painted Trillium *(Trillium undulatum)* and Dwarf Ginseng *(Panax trifolium),* and these are followed by Foamflower, Miterwort *(Mitella diphylla),* False Solomon's-seal *(Smilacina racemosa),* White Baneberry *(Actaea pachypoda),* Wood Sorrel *(Oxalis montana),* and Indian Cucumber Root *(Medeola virginiana).* Considerable areas are covered with Creeping Phlox, Canada Violet *(Viola canadensis),* or *Trillium cuneatum.* In lesser numbers are the handsome Vasey's Trillium *(T. vaseyi)* and Speckled Wood Lily *(Clintonia umbellulata).* Around the streams there are large cushions of Thyme-leaved Bluets.

As originally planned, the highway (N.C. 1127) that goes past the entrance to the Memorial Forest would have run between the Joyce Kilmer and Slickrock basins. Protests by conservationists who believed this would destroy the wilderness qualities of both brought about the abandonment of the project, and as a consequence the road ends abruptly a few miles beyond. Just before you reach that point you will see a parking pullout on the right, and directly across a sign reading Haoe Trail; the first part of this is worth investigating.

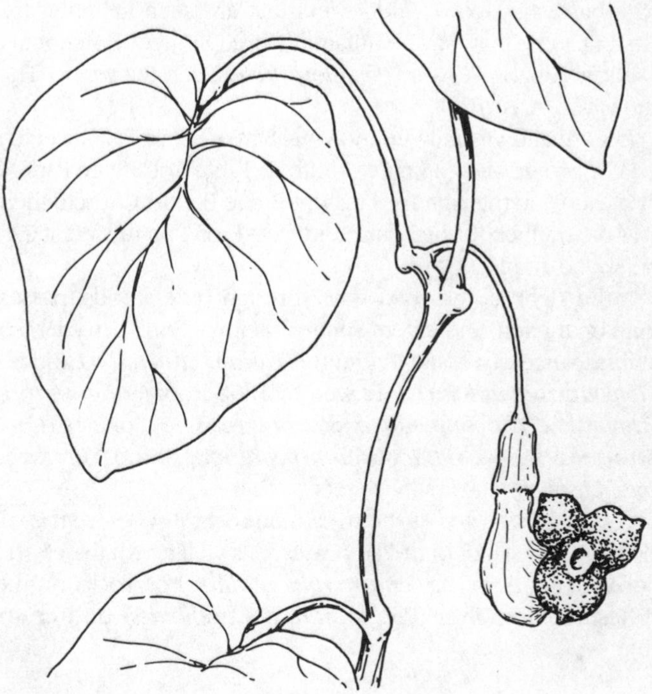

DUTCHMAN'S PIPE
Aristolochia macrophylla

The trail climbs steeply at first through a rather uninteresting dry oak woods, but as soon as it crosses to the other flank of the ridge its character changes. The first indication is Large-flowered Trillium *(Trillium grandiflorum),* then there are Yellow Mandarin, Giant Chickweed, and a rapid succession of ephemerals. Yellow Lady's Slippers *(Cypripedium calceolus)* are here, but few and hard to spot. Trees include Silverbell (which grow to enormous size here), Striped Maple *(Acer pensylvanicum),* and Mountain Magnolia *(Magnolia fraseri);* and Dutchman's Pipe Vine *(Aristolochia macrophylla)* is plentiful. Where the trail becomes rocky there are Umbrella Leaf and White Baneberry downslope and patches of Crested Dwarf Iris on the upper banks.

SOUTH CAROLINA

Only one corner of South Carolina lies in mountainous country, but it includes some extremely interesting botanical areas. One is the Carrick's Creek Trail, a beautiful interpretive walk in Table Rock State Park, on S.C. 11 north of Pickens.

Many of the hardwood cove wildflowers that are seen in other areas will be found here, but some that may be unfamiliar are Oilnut or Buffalonut *(Pyrularia pubera),* Yellow Passion Flower *(Passiflora lutea),* and the yellow Three-parted-leaf Violet *(Viola tripartita).*

More obscure and virtually unknown is Station Cove Falls, which is reached from S.C. 11 approximately 26 miles south of Table Rock State Park. Turn north on County Road 95 at the sign for Pleasant Ridge Baptist Church, then left at the fork. Park at the trailhead, which is an old wood road on the left at 2.4 miles, just before the paved road ends.

Keep to the right as the road goes through pine woods, passes an active beaver pond on the left, and enters Sumter National Forest. In early spring there is a great abundance of Sessile Trillium *(Trillium cuneatum)* and Sharp-lobed Hepatica *(Hepatica acutiloba),* followed by Bloodroot *(Sanguinaria canadensis),* Rue Anemone *(Anemonella thalictroides),* Blue Cohosh *(Caulophyllum thalictroides),* May Apple *(Podophyllum peltatum),* Shrub Yellowroot *(Xanthorhiza simplicissima),* and many others.

It is necessary to step across the stream on rocks just before the falls, but this is easily done and the sight of the lacy cascade is well worth the effort. There are colonies of Walking Fern *(Asplenium rhizophyllum)* on rocks on the opposite bank and Resurrection Fern *(Polypodium polypodioides)* on tree trunks.

GEORGIA

Nearly 1,500 miles from its beginnings in Quebec, the backbone of the Appalachian Mountains system comes to an end a mere 50 miles inside of Georgia's borders, but not before endowing that state with some of its most rugged and beautiful scenery.

Outstanding, though small, is the little-known Sosebee Cove, in Chattahoochee National Forest. Sosebee was the family name of a lifelong resident of the cove, but it might well have been named for Arthur Woody, the ranger who persuaded the Forest Service to purchase it in 1925 for its superlative stand of Tulip Trees, which, fortunately, were not cut. Although originally singled out for preservation because of its hardwood trees—including the world's largest Yellow Buckeye, a giant with a 15-foot, 10-inch girth—its present appeal as a scenic area is largely due to its wealth of spring wildflowers. It is truly a classic southern cove hardwoods forest.

Beginning at a bed of Bloodroot, a trail goes through masses of Foamflower, Carolina Spring Beauty, Giant Chickweed, Solomon's-seal *(Polygonatum biflorum),* and Downy Yellow Violet *(Viola pubescens).* There are three species of Trillium: Large-flowered, Nodding, and Wake Robin *(T. grandiflorum, T. cernuum,* and *T. erectum),* as well as Showy Orchis, Dutchman's Breeches, Blue Cohosh, and Early Meadow-rue *(Thalictrum dioicum).* Along the banks of a small brook are Trout Lily *(Erythronium americanum)* and Jack-in-the-Pulpit. Especially noteworthy is Nodding Mandarin *(Disporum maculatum),* with large purple-dotted white flowers, growing not far from the more common Yellow Mandarin *(D. lanuginosum).*

To reach Sosebee Cove, take U.S. 19 north from Dahlonega for 23 miles, then turn left on Ga. 180 for 3 miles. This route takes you past Vogel State Park, which also has pleasant walking trails. Although it lacks Sosebee Cove's variety of wildflowers, this park does have an abundance of Showy Orchis.

SHOWY ORCHIS
Galearis spectabilis

9

Mountain Balds

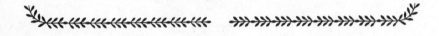

IT IS NOT possible to spend much time in the southern Appalachians without noticing that the undulating green carpet of forest is broken here and there by seemingly barren mountaintops and ridges. These are the mysterious "balds"—areas that are naturally devoid of trees where trees would be expected. They are mysterious because there seem to be no explanations of how they originated or how they are maintained.

Although the balds are confined to the higher elevations, severe weather alone is not the answer, for even loftier peaks nearby are forested all the way to the top. Mountains here would have to be more than 8,000 feet high to possess a treeline above which altitudinal factors would prevent tree growth. It has been suggested that the balds were caused by clearing for a variety of reasons by pioneers or Indians, but old accounts and legends indicate their earlier existence. Other theories attribute their formation to lightning-caused fires, ice storms, climatic changes, parasites, and other disturbances. It seems likely that when the riddle of the balds is finally solved, different answers will be seen to apply to different locations.

There are two basic types of balds, sometimes intergrading. Heath balds, or laurel slicks as they are called locally, often occur on rocky sites and are dominated by low-growing shrubs. As the name suggests, members of the Heath family (Ericaceae) prevail, and Catawba Rhododendron *(Rhododendron catawbiense)* is by far the best known of these. Grass balds are meadowlike and have a thick cover of long grasses, notably Allegheny Fly-back *(Danthonia compressa)*. They are also host to many herbaceous wildflowers.

ROAN MOUNTAIN

There can be no doubt that Roan Mountain is the most exciting of all the balds. Certainly it is the most publicized as well owing to the Rhododendron Festival held during the second half of June every year.

The focus of attention is not on the town but on the mountain itself or, more precisely, Carvers Gap on the boundary line where Tenn. 143 meets N.C. 261 and is joined for good measure by the Appalachian Trail. To see the most spectacular displays of Catawba Rhododendron in the world, start out on Forest Service Road 130, which goes westward from this same intersection. There are three parking lots along the route, each with access to a trail; here you can walk or enjoy a picnic lunch in a mazelike garden formed by walls of these purple-flowered shrubs towering overhead. It is difficult indeed to believe that these are not cultivated, but except for minimal management to prevent intrusion by objectionable vegetation, they are entirely natural.

The road ends in a wide loop, with yet another parking space for the short trail through a boreal forest to Roan High Bluff, a rocky overlook 6,267 feet above sea level. Michaux's Saxifrage *(Saxifraga michauxii)* is extremely abundant along this path. A great many wildflowers will be found along the roadsides, especially around this circle. The most significant is Spreading Avens *(Geum radiatum),* a large-flowered yellow species closely related to the *Geum peckii* of New England's alpine meadows.

Returning to Carvers Gap and walking in the opposite direction along the Appalachian Trail, you ascend a series of steps to Round Bald. Here on this grassy bald the Catawba is scattered along with an occasional Flame Azalea *(Rhododendron calendulaceum),* and clumps of Green Alder *(Alnus crispa)* and a few Hawthorns (*Crataegus* sp.) just about round out the woody vegetation. Blankets of small white flowers turn out to be sheets of Greenland Sandwort *(Arenaria groenlandica)* and of Wine-leaved Cinquefoil *(Potentilla tridentata),* two more plants that are better known to those familiar with the northern mountains. And there are patches of the truncate-leaved Robbins' Ragwort *(Senecio robbinsii),* a rare disjunct found nowhere else south of upper New York State. Much more at home are Mountain St. John's-wort *(Hypericum mitchellianum),* Tassel-rue *(Trautvetteria carolinensis),* and Filmy Angelica *(Angelica triquinata).*

If you are here in early July you will see all of these, but you will also be treated to the delightful red-orange Roan Lily *(Lilium grayi)* where it was discovered (see next page for illustration). Although listed as threatened in the state, it seems to be doing well here, but in some other places like the Great Smokies the bulbs are being rooted up by wild boars. These animals, which were imported from Europe, escaped from a game preserve in the 1920s and have become a serious problem.

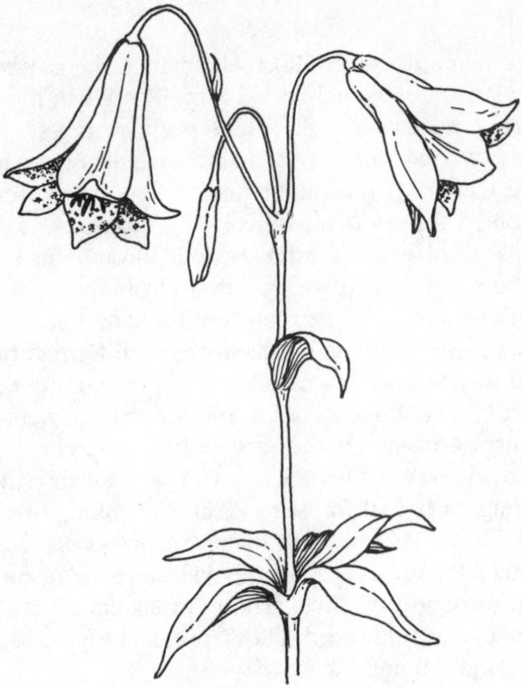

ROAN LILY
Lilium grayi

CRAGGY PINNACLE

Two of the trails in the Craggy Gardens area of the Blue Ridge Parkway that provide access to balds were described in Chapter 7, on northern hardwoods forests. Another popular trail ascends Craggy Pinnacle from the parking lot immediately north of the tunnel at mile 364.4.

All of these balds afford an excellent alternate to Roan Mountain for seeing masses of Catawba Rhododendron somewhat earlier in June, as well as other ericaceous shrubs: Minnie-bush *(Menziesia pilosa),* Mountain Laurel *(Kalmia latifolia),* and several Blueberries (*Vaccinium* spp.). The grassy sections are also good places for Wine-leaved Cinquefoil.

OLD BALD

Although it is not identified by any sign, Old Bald is easily reached by parking on the Blue Ridge Parkway shoulder at mile 433.8 and entering the gated path. This parallels the parkway, with a wire fence on the right partially hidden by immature Beeches *(Fagus grandifolia),* Hawthorns *(Crataegus flabellata),* and Elderberry *(Sambucus canadensis).*

Just beyond another gate the fence turns right and the trail forks. The right-hand prong leads uphill to the grassy top of the bald proper. Scattered throughout is Wine-leaved Cinquefoil, often growing amid silvery cushions of Reindeer Lichen (*Cladonia* sp.), which provide an attractive background for the red leaves in autumn. By proceeding straight ahead at the fork, you pass through some shrubby undergrowth, then emerge into a small clear area where there is a small population of the uncommon Frostweed *(Helianthemum bicknellii).*

Although an abundance of Trailing Arbutus *(Epigaea repens)* alone is enough to recommend Old Bald for a spring visit, the best times probably are late summer and early fall. The white-flowered Silverrod *(Solidago bicolor)* is everywhere, and of course there are Asters, including *A. pilosus, A. paternus,* and *A. linariifolius.* Pinks and purples are supplied by Phlox *(Phlox carolina)* and Purple Bergamot *(Monarda media).* Both Striped Gentian *(Gentiana decora)* and Stiff Gentian *(G. quinquefolia)* are in evidence; in the southern mountains the latter achieves a particularly deep blue color. Orchids are represented by the fragrant Nodding Ladies' Tresses *(Spiranthes cernua).*

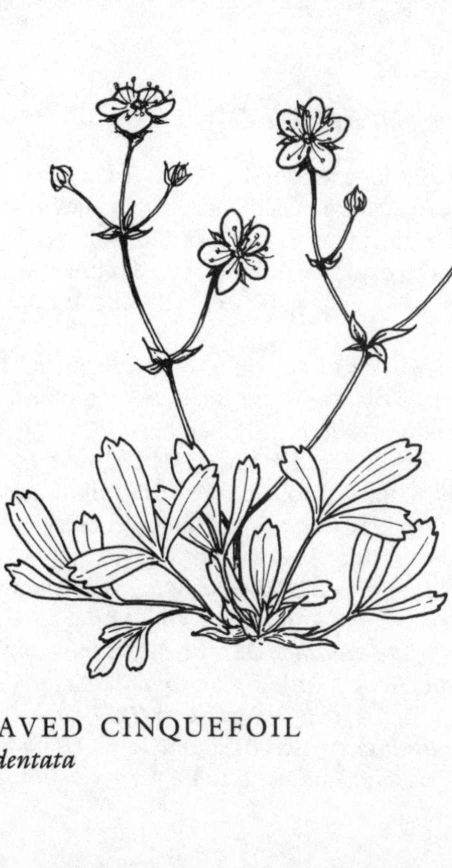

WINE-LEAVED CINQUEFOIL
Potentilla tridentata

In August and September you can pick blueberries to your heart's content. You may meet some Cherokee Indian families doing the same thing, which is a reminder that about 25 miles to the west the parkway enters the Qualla Reservation. The most plentiful Blueberry is a low-bush species, *Vaccinium vacillans,* the foliage of which has a delightful way of turning deep rose while the powdery-blue berries are still on the bush.

ANDREWS BALD

Much has been written about the spectacular displays of Flame Azaleas in the Great Smoky Mountains, and probably few can surpass those on Gregory Bald. For hikers who prefer a much shorter trail, however, the 2-miler from the Clingman's Dome parking area to Andrews Bald offers a tempting alternative.

After an initial passage that is rocky underfoot, this becomes a pleasant walk through a deep, mossy Fraser Fir and Red Spruce forest. The trail eventually opens out onto the bald itself, which is a wide meadow of grasses and sedges affording beautiful views toward the south. In June both Flame Azalea and Catawba Rhododendron are in bloom, and these are followed by Filmy Angelica and various composites. Striped Gentians finish the flowering season, persisting into October.

WEST VIRGINIA

A number of bare mountaintops are found in West Virginia, and there is some question whether these are "natural" balds or whether they once had a forest cover that was destroyed over and over again by a succession of fires. In any case they are interesting places to explore, and the obvious choice would be Spruce Knob in Pendleton County, which at 4,860 feet is the highest point in the state, yet is accessible by road.

Much of the summit is what is locally called a "heath barren" because of the great number of ericaceous plants. These include Minnie-bush, Black Huckleberry, and several *Vacciniums,* including Southern Mountain Cranberry *(V. erythrocarpon)* near its northernmost limit here. Where there are trees they are stunted Red Spruce, Yellow Birch, Mountain Ash, and the like.

Those familiar with the balds to the south will be struck by the differences in the flora of this region. Spruce Knob is just a few miles too far north for Catawba Rhododendron, but its place is taken by the beautiful, very fragrant Rose Azalea (*Rhododendron prinophyllum,* ex. *R. roseum*). One of the southernmost populations of Bunchberry *(Cornus canadensis)* is on the summit. *Ilex ambigua* var. *montana* is here, but so is *Nemopanthus mucronata,* which is also known as Mountain Holly. Common herbs include Fireweed *(Epilobium angustifolium),* Pearly Everlasting *(Anaphalis margaritacea),* and Wild Bleeding-heart *(Dicentra eximia),* the last especially abundant along the entrance road.

10

Southern Pinelands

As we move out of the southern Appalachians, through the piedmont, and onto the vast coastal plain, we leave behind a number of pines: Eastern White, Pitch, Table Mountain, and gradually, Virginia or Scrub Pine. Their places are taken by several others (all so-called hard pines), and although we see more and more large, uniform expanses of single species, others possess very limited adaptability and are found only in small, highly specialized habitats.

During colonial times the most widespread and useful of these trees was Longleaf Pine *(Pinus palustris).* But because it was heavily exploited for its lumber and for naval stores (turpentine, pitch, and resin) and was replaced by agricultural enterprises (in Florida's citrus belt, for example), its distribution has been greatly reduced.

Longleaf Pine is an important constituent of pine savannas, where well-spaced scatterings of the trees overtop beds of Wiregrass *(Aristida stricta)* in which a wealth of interesting and often rare herbaceous flowers flourish. A slow-growing species, Longleaf Pine spends up to the first 10 years of its life in what is called the "grass stage"; that is, as a short, stubby seedling clothed in plumes of long needles. During this period the root system is being rapidly developed, but the stem does not elongate. Of all the pines, Longleaf is especially dependent upon frequent fires to discourage competitive woody plants and periodically dispose of accumulated leaf litter. The thick bark of the mature trees and the heavy layer of expendable needles on the "grass" seedlings afford ample protection.

Two other common species have ranges that extend farther inland and may be found mixed with hardwoods. They are Shortleaf Pine *(P. echinata),* a valuable source of timber, and Loblolly *(P. taeda),* the fastest growing of all the southern pines and therefore somewhat inferior as lumber. In a limited portion of their ranges on the coastal plain these two species are sometimes joined by Spruce Pine *(P. glabra),* a tree of no appreciable economic importance.

Slash Pine *(P. elliottii)* is the only species that is found from one end of Florida to the other, and because of its rapid growth it is being widely used for reforestation. It is an important source of naval stores, surpassing even Longleaf in the production of crude resin; in fact, its common name refers to the "face" that is cut into the bark to obtain the sap. In addition, it is harvested for pulpwood and for its termite-resistant lumber.

A particular kind of pine habitat called "rockland" forms an arm starting from near Fort Lauderdale, running through Miami, then curving southwest into the heart of Everglades National Park and emerging in the Keys. The pines growing on this limestone ridge are differentiated as *P. elliottii* var. *densa* and are commonly referred to as South Florida or Dade County Slash Pine because of their limited distribution. They have been largely depleted, not only by commercial exploitation but by the clearing of the land on which they grow (which being slightly elevated is desirable for agriculture and housing) and by human interference with the natural incidence of fires.

A small bushy tree, Sand Pine *(P. clausa)* grows in poor soils, principally on ancient dunes in northern and coastal Florida. In this desertlike environment it is associated with many other unusual plants in forming sand pine scrub, a community peculiar to the state.

Pond Pine *(P. serotina),* on the other hand, needs a moist habitat. It is the dominant conifer, along with Pond Cypress *(Taxodium ascendens),* in many of the Carolina "bays" that, because of their unique character, are being accorded a chapter of their own.

Two members of the Palm family are a prominent feature of the more southern of the pinelands. One is Cabbage Palm *(Sabal palmetto),* the classic native palm of the American South. The word "cabbage" refers to the terminal bud, which gourmets glorify as "heart-of-palm" but natives are content to call "swamp cabbage." It is considered a culinary delicacy, but its removal is fatal to the tree. The other is the fan-shaped Saw Palmetto *(Serenoa repens),* by far the most abundant native palm. It has a shrubby aspect as its stem usually protrudes only a short distance above ground. Sawtoothlike spines on the edges of the leafstalks account for the common name.

THE CAROLINA SANDHILLS

Mention the sandhills of the southeastern coastal plain and you are not likely to elicit any particular response, but throw out the names Pinehurst, Southern Pines, and Pine Bluff and you will bring every golfer within earshot to attention.

These flat sandy ridges were once clothed in extensive forests of Longleaf Pine and were an important source of naval stores, but late in the 19th century it was recognized that they had another appeal. A cluster of resort communities sprang up in the south-central part of North Carolina, and the region quickly became a fashionable vacation area, with golf and riding the principal sports as they are today.

A remnant of these pinelands is contained within the Weymouth Woods Sandhills Nature Preserve, located 3 miles from U.S. 1 south of Southern Pines. Some typical wildflowers along the trails are Goat's Rue *(Tephrosia virginiana),* Samson Snakeroot *(Psoralea psoralioides),* and Whorled Coreopsis *(Coreopsis verticillata),* while Trumpets *(Sarracenia flava)* and Sundews inhabit the wetter areas.

An extension of these sandhills into South Carolina takes in the Carolina Sandhills National Wildlife Refuge, which has both roads and trails. The entrance is located 4 miles north of McBee on U.S. 1.

Certainly the most unusual plant here must be the Small-leaved Pyxie Moss (*Pyxidanthera barbulata* var. *brevifolia*). This forms spreading mats spangled with exceptionally pretty flowers in early spring—a scaled-down form of the typical species that we observed in the New Jersey Pine Barrens. Again there are several diverse habitats, including Turkey Oak *(Quercus laevis)* woods, pocosins, and bogs.

As might be expected, this is also an excellent place for observing wildlife, and the old-growth Longleaf Pines provide nesting sites for large numbers of red-cockaded woodpeckers. The eastern fox squirrel and the beautiful pine barrens tree frog are among the many other residents.

GREEN SWAMP

North Carolina's Green Swamp lies southwest of Wilmington in Brunswick County. Several other states also have "Green Swamps," but in this case the name is not descriptive but derives from one John Green, who was granted title to the land by the English Crown prior to the American Revolution.

The Nature Conservancy has under its ownership and protection a 13,850-acre portion of the swamp, which really includes grassy pine savannas occupying scattered "islands," which prove to be low sandy ridges, as well as slightly elevated acidic bogs dominated by evergreen shrubs and Pond Pine. The latter are called "pocosins," from the Indian word meaning "swamp on a hill."

Field trips under the guidance of competent leaders are offered by the conservancy and, of course, are highly recommended. A brief but exciting introduction to these plant communities can be had, however, by pulling off on the east side of N.C. 211 at a borrow pit north of the town of Supply. Immediately to the right is a typical parklike savanna with widely spaced Pines and a mixture of grasses underfoot, notably Wiregrass and Toothache Grass *(Ctenium aromaticum),* studded with wildflowers.

TRUMPETS
Sarracenia flava

One of the first plants to be noticed is Venus' Flytrap *(Dionaea muscipula),* with its reddish opened leaves facing upward and attractive white flowers topping off a rather tall scape. These savannas and the nearly barren sands around the swamp's borders are home to many carnivorous plants (every one of the 14 species native to the state has been found in Green Swamp). In this immediate area you will see the striking yellow Trumpets *(Sarracenia flava)* and other Pitcher Plants such as Hooded *(S. minor),* Red *(S. rubra),* and Northern *(S. purpurea);* Pink and Spatulate-leaved Sundews (*Drosera capillaris* and *D. intermedia*); and Butterworts (*Pinguicula* spp.).

Other inhabitants include terrestrial orchids: Grass Pink *(Calopogon tuberosus)* with its paler relative *(C. pallidus),* Rose Pogonia *(Pogonia ophioglossoides),* the elegant Spreading Pogonia *(Cleistes divaricata),* and Ladies' Tresses *(Spiranthes).*

There are rare plants, too, like Carolina Grass-of-Parnassus *(Parnassia caroliniana).* A small water-filled depression has the white Spoon Flower *(Peltandra sagittifolia),* and near the edge of the savanna we see the handsome Rough-leaf Loosestrife *(Lysimachia asperulaefolia)*—both of them endangered in the state. Obviously the list could go on and on through more prosaic categories of plants.

A short walk through the savanna brings you to a boardwalk that bridges an intervening pocosin, and here the vegetation changes abruptly. You can pick out such trees as Pond Pine, Pond Cypress, Black Gum *(Nyssa sylvatica),* and Sweet

Bay *(Magnolia virginiana),* but it is the shrubs that make up this impenetrable thicket. These include Sweet Pepperbush *(Clethra alnifolia),* Ti-Ti *(Cyrilla racemiflora),* Southern Bayberry *(Myrica cerifera),* Red Bay *(Persea borbonia),* and several species of *Ilex,* among them Sweet Gallberry and Myrtle-leaf Holly (*I. coriacea* and *I. cassine* var. *myrtifolia*). As might be expected, a large number are in the Heath family: Swamp Fetterbush *(Leucothoe racemosa)* and several species of *Lyonia* and *Vaccinium* among them.

The flora on the rim of the borrow pit partakes of that in the savannas but offers others like the startling Yellow Meadow Beauty *(Rhexia lutea),* an abundance of Redroot *(Lacnanthes caroliniana),* and several *Polygalas.* In the drier sands we find the little Roseling *(Tradescantia rosea)* and the even smaller *Wahlenbergia marginata.*

VENUS' FLYTRAP

Situated on a barrier island near Cape Fear, Carolina Beach State Park encompasses a number of coastal habitats worth exploring, but most noteworthy among its attractions is the presence of Venus' Flytrap in great numbers. Although this plant is not rare, as many suppose, it grows nowhere in the world except in southeastern North Carolina and a few places along the South Carolina shore.

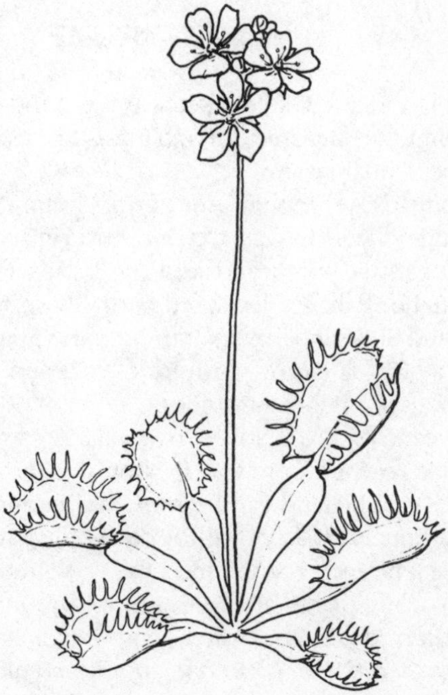

VENUS' FLYTRAP
Dionaea muscipula

It is difficult to question Charles Darwin's unequivocal description of Venus' Flytrap as "the most wonderful plant in the world" when we examine the ingenious apparatus with which it captures insect prey to supplement its diet. As with most carnivorous plants, it is the leaves that perform this function. But what fascinates us is that the process is not passive as it is in the Pitcher Plants, slow as in the Sundews, or hidden as in the Bladderworts, but takes place in full view as a fast-acting "snap trap." We can even test its action with a sprig of grass.

The winged leafstalks each terminate in a hinged pair of blades, which bear reddish nectariferous glands on the inner surfaces and a row of long spines on the margins. There are three trigger hairs on either side of the midrib, and when an insect touches down on these the two lobes of the leaf snap together quickly, and the spines interlock to form a cage imprisoning the prey. This mechanism is incredibly fine-tuned. For the trap to be sprung, one trigger hair must be touched twice or two separate hairs touched within a certain time span; the plant is even able to differentiate between an insect and the impact of rain, dust, or wind, and to some extent between an insect that constitutes a worthwhile meal and one that is too insignificant. After the digestible parts have been consumed, the leaf unfolds, ready for a new victim.

In addition to its other paths, Carolina Beach State Park has a half-mile Flytrap Trail. The entrance is located 15 miles south of Wilmington off U.S. 421.

OKEFENOKEE

To the Seminole Indians, Okefenokee was the "Land of the Trembling Earth." To Walt Kelly's comic strip character Pogo it was home. And to just about everyone it is the quintessential swamp.

Despite long usage, though, the term "swamp" is not appropriate when applied to Okefenokee. This vast wetland, covering something like a half-million acres in southeastern Georgia and adjacent Florida, is actually an enormous peat bog, slightly higher than sea level, fed partly by springs, and drained slowly by the St. Marys and Suwannee rivers into both the Atlantic Ocean and the Gulf of Mexico. Much of it is primitive, trackless wilderness, and cannot safely be explored without a competent guide.

Ninety percent of Okefenokee is contained within the National Wildlife Refuge, which is reached from Ga. 23 about 7 miles southwest of Folkston, Ga. Here two types of pineland can be observed very readily.

Botanically minded visitors will be tempted to linger over the wide shoulders lining the entrance road, which they will recognize as prolific savanna habitat. Carnivorous plants are represented by blue and yellow Butterworts (*Pinguicula caerulea* and *P. lutea*) and Parrot and Hooded Pitcher Plants (*Sarracenia psittacina* and *S. minor*), and the orchids by Many-flowered Grass

Pink *(Calopogon multiflorus).* Milkworts include Orange Bachelor-button *(Polygala lutea)* and the bright yellow Candyroot *(P. nana).* White Narrow-leaved Violets (*Viola lanceolata* var. *vittata*) are prolific in the wet spots.

A short distance inside the refuge is the Canal Diggers Trail, which winds through Longleaf Pine uplands where a futile attempt was made in the 1890s to drain the swamp. Here and along the sandy roadsides a very different assortment of flowering herbs will be encountered, including several that are frequently seen in dry southern woods: Prickly Pear Cactus *(Opuntia compressa),* Spurge Nettle *(Cnidoscolus stimulosus),* and Roseling. Others that may be less familiar are Blackroot *(Pterocaulon pycnostachyum),* a yellow Colic Root *(Aletris lutea),* and Pygmy Pawpaw *(Asimina pygmaea).* Moister areas contain the white form of Maryland Meadow Beauty *(Rhexia mariana)* and an occasional Spreading Pogonia orchid.

It is possible, of course, to sample other environments from this point as well. There are several trails, a 4,000-foot boardwalk, and a restored island homestead. Motorboats and canoes can be rented for modest excursions on the Suwannee Canal, where Cypresses arch over the mirrorlike waters and alligators seem to be everywhere (Okefenokee's population of these reptiles is estimated at more than twelve thousand). From the main waterway you can enter some of the "prairies," which are really open lakes adrift with Water Lilies *(Nymphaea odorata),* Spatterdock *(Nuphar macrophyllum),* Floating Bladderwort *(Utricularia inflata),* and Golden Club *(Orontium aquaticum)* in incredible numbers.

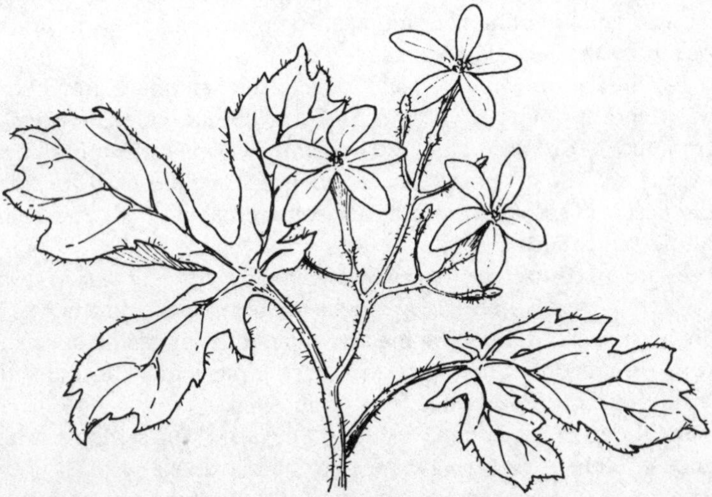

SPURGE NETTLE
Cnidoscolus stimulosus

GOLDEN CLUB
Orontium aquaticum

SAND PINE SCRUB

Florida's Sand Pine Scrub—which really ought to have a more inspiring name—is not only unique to a state already famous for its unusual flora, but is the oldest on a peninsula that is itself relatively young. Its habitat is a foundation of limestone formed when Florida was under the sea, overlaid with glistening white "sugar sand" of mineral origin washed down from the north, and eventually exposed by the receding waters.

By far the largest area of sand pine scrub anywhere lies between the Oklawaha and St. Johns rivers, much of it in Ocala National Forest. Near its western limits, at Cross Creek, is the home of Marjorie Kinnan Rawlings, author of *The Yearling,* who gave many readers their first and most vivid descriptions of the "Big Scrub." Outlying remnants are scattered along both coasts and in the central highlands farther south.

These are, of course, the places where Sand Pine *(Pinus clausa)* grows. This small, scrawny tree is not as dependent upon periodic burnings as Longleaf Pine; in fact, hot crown fires leaping across the treetops can be devastating to it. Nevertheless, it is only when fires do bring the pines into check that the scrub component of these forests becomes established.

One of the most remarkable features of the sand pine scrub flora is the high incidence of endemics. It has been estimated that about one-half of the plants are to be found nowhere else, which is several times the rate of endemism for Florida taken as a whole. Many others with slightly less stringent requirements occur in sandy pine flatwoods as well.

Very prominent among the "indicator" plants are three small oaks: Chapman's Oak *(Quercus chapmanii),* Myrtle Oak *(Q. myrtifolia),* and Scrub Live Oak *(Q. geminata).* Other species, some of them much less common, include Silkbay *(Persea humilis),* Pygmy Fringe Tree *(Chionanthus pygmaeus),* Bigfruit Devilwood *(Osmanthus megacarpus),* Pawpaw *(Asimina reticulata),* Scrub Mint (*Conradina* spp.), Scrub Clover *(Dalea feayi),* and Wireweed *(Polygonella myriophylla).*

Ocala National Forest

One of the principal attractions in Ocala National Forest is its crystal clear springs, several of which have public recreation areas associated with them where there is swimming and boating. One favorite is Juniper Springs, which includes a Sand Pine forest; it is located 26 miles east of Ocala on S.R. 40. Walking along the footpaths here, one is especially aware of the more primitive plant forms, like Coontie *(Zamia pumila),* and clusters of Resurrection Fern *(Polypody polypodioides)* on massive oak limbs that appear to have been painted with the bright pink Blanket Lichen *(Herpothalion sanguineum).* Fungi emerge from the seemingly dry edges of the trail, among them the red Columnar Stinkhorn *(Clathrus columnatus).* In wetter areas Southern Red Cedar *(Juniperus silicicola)* may be seen.

A little to the southeast is Alexander Springs, where 78 million gallons of water bubble up each day, and an interpretive trail winds through a humid environment. The recreation area is on County Road 445, east of S.R. 19.

Clearwater Lake is on the southern edge of the National Forest, on S.R. 42 about 16 miles west of Deland. Here you might look for Dwarf Sundew *(Drosera brevifolia)* along the shore and the little prostrate white "Bluet," *Hedyotis procumbens,* in the pine woods.

Developed areas such as these may prove more satisfactory than the network of hiking trails elsewhere in the national forest because of timber management and production activities—unless, of course, you intend to camp. Not far from its boundaries are two state parks that also can be recommended. One is Wekiwa Springs (north of Apopka), where both a scenic drive and a self-guiding nature trail take you through sand pine scrub and other types of vegetation. The other is Blue Spring (south of Deland); here the major attraction is the gathering of manatees during the winter and early spring.

Jonathan Dickinson State Park

Another excellent example of sand pine scrub is located in Jonathan Dickinson State Park, on the west side of U.S. 1 about 6 miles north of Jupiter. (If you are approaching from the direction of Stuart, you will notice whole forests along the highway where every tree leans permanently landward—the result of hurricane winds in the past.)

The scrub forest is just inside the park entrance along what is called the Hobe Mountain Trail. The name might come as a surprise in view of the flat terrain, but anyone who may be apprehensive about the altitude needs only to be reminded that this is also the state in which people wear T-shirts proclaiming "I Climbed Mount Dora" (a feat roughly equivalent to going upstairs).

Almost immediately you are conscious of the three stunted oaks previously mentioned, and of course Sand Pine. The inevitable Saw Palmetto is joined by another with an underground spiral stem; this is Scrub Palmetto *(Sabal etonia).* A sand-loving companion of these is a Prickly Pear Cactus, *Opuntia humifusa.*

Rosemary is here, too. Not the cooking herb, but *Ceratiola ericoides,* a dense shrub with many ascending branches bristling with needlelike leaves in four ranks. Among the wildflowers here are the showy Rattlebox *(Crotalaria spectabilis),* Golden Aster *(Heterotheca scabrella),* Sandhill Wireweed (*Polygonella fimbriata* var. *robusta*), and Palafoxia *(Palafoxia feayi).*

This park also has a pleasant path (the Kitching Creek Trail) through pine flatwoods, where Slash Pine *(Pinus elliottii)* and Turkey Oak *(Quercus laevis)* are the principal representatives of their genera.

Several evergreen shrubs will be noted. Probably the most plentiful is *Lyonia lucida,* one of several heaths that have been given the name of Fetterbush, and which bears deep pink blossoms. In the same family is the Shiny Blueberry, *Vaccinium myrsinites.* Gallberry, or Inkberry *(Ilex glabra),* is a holly with distinctive black berries that can be found all the way up the Atlantic coast into Canada.

A member of the Mint family that is restricted to the lower half of Florida but can be expected in almost any dry pine woods here is *Piloblephis rigida.* Compact racemes of purplish pink flowers top the branches of this low herb, which are covered with very numerous short, linear leaves. It shares the common name of Pennyroyal with *Hedeoma pulegioides,* but this should cause no real difficulty since their ranges do not overlap. Another attractive little plant is *Hedyotis procumbens,* sometimes called Innocence. Much more flamboyant is the Tickseed *Coreopsis leavenworthii,* its dark brown central disk contrasting sharply with the butter yellow ray florets.

Wilson Creek, which is crossed and recrossed by the trail, is almost negligible in size but accounts for an abrupt change in vegetation. Suddenly there are Wax Myrtle, Swamp Fern, Willows, and even a few small Cypresses. In one place along the creek, trees and shrubs are completely hidden by masses of Japanese Climbing Fern *(Lygodium japonicum),* obviously an escape from cultivation. Fortunately, at one of the bridges it is possible to examine this beautiful fern at close range.

LONG PINE KEY

One of the few places in mainland Florida where South Florida Slash Pine (*Pinus elliottii* var. *densa*) is being accorded permanent protection is at Long Pine Key in Everglades National Park, not far from the main entrance.

The predominant accompaniment is Saw Palmetto, but here this is joined by Silver Palm *(Coccothrinax argentata),* which has fronds that are silvery white beneath and unarmed leafstalks. Silver Palm is also plentiful in the Bahamas, and at the little settlement of Red Bay on Andros Island the descendants of former slaves of the Seminole Indians weave durable baskets, and even water pitchers, from the fronds.

Another plant of these pine rocklands is Short-leaved Fig *(Ficus laevigata).* It begins life as an epiphyte and sends aerial roots down to the soil, and because of this has been called Wild Banyan. It does not, however, suffocate its host as does the Strangler Fig.

An unusual plant that formerly was much more common in Florida's pine woods is Coontie *(Zamia pumila).* This is one of the cycads, a family of primitive tropical lineage. Coontie has dark green fernlike foliage and its stem is underground, so that it looks a little like the top of a buried palm tree. Stout brown cones that produce bright orange seeds are a conspicuous feature. For centuries Indian tribes made an edible starch from the roots, and as a result of this Coontie joined the long list of those to which the name Arrowroot has been loosely applied.

COONTIE
Zamia pumila

11

The Mysterious Carolina Bays

IMAGINE YOURSELF in a hot air balloon floating high above the southern coastal plain. Far beneath the drifting wisps of white clouds, the land appears as a flat mosaic of fields, seemingly devoid of any other features.

Dropping down for a better look, however, you discern a haphazard pattern of dark elliptical patches superimposed on the landscape. They vary greatly in size and are randomly distributed, sometimes touching and even overlapping, yet they are surprisingly uniform in their orientation, each having its long axis pointed almost exactly southeast and northwest. Curious now, you go still lower and find that they are shallow depressions each surrounded by a low rim of sand.

What you are looking at—and seeing from the best vantage point, aloft—is one of the earth's most fascinating and mystifying terrestrial phenomena. They are called the "Carolina bays"; the first part of the term reflects the fact that although there are hundreds of thousands of them between Maryland and Florida, the greatest concentrations are in North and South Carolina.

To understand the use of the word "bay" we need to go back to 1700, when European explorers became aware that many of the freshwater wetlands on the coastal plain were populated by assortments of broad-leaved evergreens. Upon closer examination they noticed that several of the common species had aromatic leaves like those of the bay trees back home, and this led to the name "bay swamp." It is interesting to note the familiar names by which some of the trees and shrubs are known today: Red Bay *(Persea borbonia),* Loblolly Bay *(Gordonia lasianthus),* and Sweet Bay *(Magnolia virginiana).*

The origin of these formations is very much a mystery. Once it was discovered that they had a common geometric configuration, several theories were advanced—a meteorite shower, soil solution, artesian springs, and wind scour were some—but none has been fully accepted as a scientific explanation.

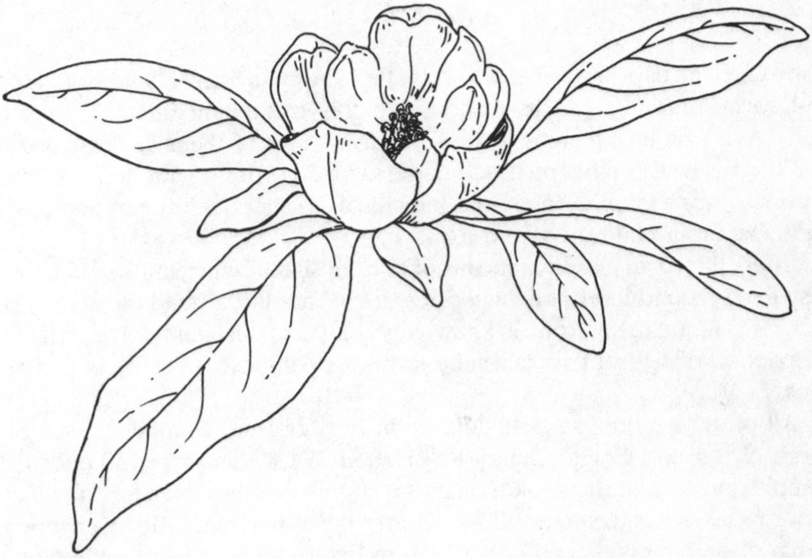

SWEET BAY
Magnolia virginiana

It is assumed that all of the Carolina bays were lakes at one time, but most are now on their way to becoming peat bogs through normal plant succession. This would seem a natural consequence of their drainage, which is poor because of the flatness of the land and a relatively impervious subsoil, yet is often sufficient to prevent surface water from accumulating permanently. The result is a soil that is soggy much of the time but dries out occasionally, and in which decomposition is incomplete—the kind most favored by acid-loving shrubs like the heaths.

This is a generalization, though, and the vegetation in and around individual bays varies greatly depending in part upon size, depth, and substrate. It runs the gamut from grasslands, marshes, and bogs to pocosins, cypress swamps, and open lakes.

Because the pocosins (we long ago reverted to the use of the original name) are dominated by shrubs, it is not surprising to find the Heath family represented by numerous species. Among them are Fetterbush *(Lyonia lucida),* Coastal Sweetbells *(Leucothoe axillaris),* Honeycup *(Zenobia pulverulenta),* Swamp Azalea *(Rhododendron viscosum),* and the less common Leatherleaf *(Chamaedaphne calyculata).* Rarest of all is White Wicky *(Kalmia cuneata),* an endemic related to Sheep Laurel but bearing a cylindrical raceme of white flowers.

Two other rarities were recently discovered in North Carolina bays: *Lobelia boykinii* and *Oxypolis canbyi.*

FIVE BAYS

Many Carolina bays have already succumbed to draining and conversion to corn and soybean farms, and in response to this continuing threat The Nature Conservancy is taking steps to preserve a number of them through outright purchase as well as other methods. At the same time, this organization is making it possible for groups of interested individuals to experience them firsthand by participating in conducted field trips.

These afford an excellent means of *finding* significant examples (not always easy from ground level), hiking *through* them (not just around the edges), and *learning* about them from a knowledgeable naturalist-guide. The following description of several bays takes the form of a composite account of two such trips.

All of the bays on our schedule lie in North Carolina's inner coastal plain south of the Fort Bragg Military Reservation. We began in the McIntosh Bay complex on a route that took us across a large expanse of meadow dominated by a species of Cutgrass *(Leersia hexandra).* It was mid-June, and this happened to have been an exceptionally wet year, so before long we were wading rather than walking, but on a return trip in a dry year there was not even a hint of moisture here. On the far side, a low sandy ridge had an abundance of Redroot (*Lacnanthes caroliniana*) foliage (it would not flower until later) and tiny rosettes of Pink Sundew (*Drosera capillaris*).

Down over the ridge the transition to swampland was abrupt. The water was shallow and dark in the welcome shade of Pond Cypress *(Taxodium ascendens)* and some Pond Pine *(Pinus serotina)* and Black Gum *(Nyssa sylvatica).* Soon we were in an area where the surface was dotted with bright yellow as far as could be seen; these turned out to be the flowers of a Bladderwort, *Utricularia fibrosa,* thrust above the water on wiry stems. A pleasant surprise was the discovery that a few plants with delicate blue flowers were the rare *Lobelia boykinii.*

Turning to leave the bay by a different route, we traversed a section of Pond Pine where the trees were being cut, and here we found some Slender Arrowhead *(Sagittaria teres)* together with less showy plants like Pipewort (*Eriocaulon* sp.) and Beak Rush *(Rhynchospora inundata).*

The second bay we entered was Antioch Church Bay, and here another bladderwort claimed our attention. This was *Utricularia inflata,* called Floating Bladderwort because of the inflated leaves that radiate like the spokes of a wheel and support the erect flower stalk. These arms terminate in masses of tiny immersed bladders that trap aquatic insects as the plant drifts almost imperceptibly, literally trolling for its food. Another yellow flower, even more plentiful, was Tall Milkwort *(Polygala cymosa).* Many plants of the Stinking Marsh Fleabane *(Pluchea foetida)* were in bud, and some seed capsules from the preceding year were identified as belonging to Awned Meadow Beauty *(Rhexia aristosa).*

This bay was also revisited during a dry period. The Bladderwort was nowhere to be seen, of course, but the Tall Milkwort was even more abundant

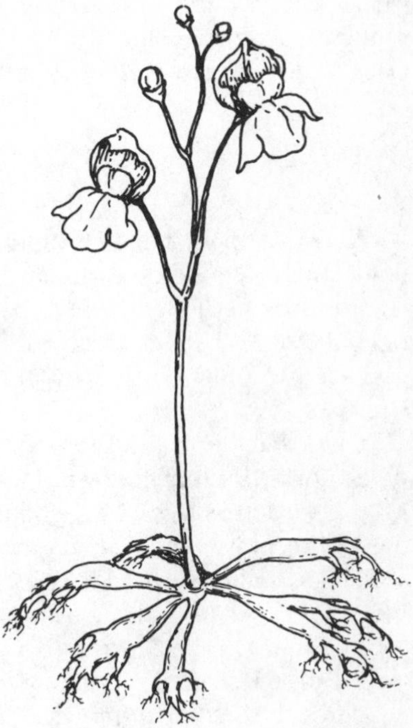

FLOATING BLADDERWORT
Utricularia inflata

than before, and the absence of deep water enabled us to enjoy the attractive little pink composite *Sclerolepis uniflora.*

Goose Pond Bay, which we came to next, held a little less interest in terms of flowering plants, but this was compensated for by ample numbers of the rare Sarvis Holly *(Ilex amelanchier),* which at the time bore immature fruit. On the rim of the bay, which borders on the road at the point where we entered, were the white-flowered form of Pale Meadow Beauty *(Rhexia mariana),* the brilliant heads of Orange Milkwort *(Polygala lutea),* Nuttall's Lobelia *(Lobelia nuttallii),* and Bog Buttons *(Lachnocaulon anceps).*

The last stop, known as Pretty Pond Bay, must be reported more briefly. Just a few feet from the edge, amid the beautiful emerald foliage of the Pond Cypresses, the water was up to our armpits, and with camera held overhead in one hand and wallet in the other the possibility of tripping or stepping into an even deeper hole was enough to discourage further exploration.

On the second trip a stop was made at Dunahoe Bay, which is a Pond Cypress *(Taxodium ascendens)* swamp and is perpetually flooded. Near the shore there were rafts of white Water Lilies *(Nymphaea odorata)* interspersed with tiny, vividly colored Purple Bladderwort *(Utricularia purpurea).* The banks were a tangle of vegetation, notably Swamp Loosestrife *(Decodon verticillatus),* with Fetterbush *(Leucothoe racemosa),* Virginia Willow *(Itea virginica),* and Netted

Chain Fern *(Woodwardia areolata)* in evidence. The approach was along a sandy road with Trailing Morning Glory *(Bonamia patens)* and the extremely diffuse Wire Plant *(Stipulicida setacea)* among the plants in flower.

EXPLORING ON YOUR OWN

It is possible, of course, to explore some Carolina bays without the benefit of a guide, but most of those with developed public access have lakes as their main feature and offer little besides a boat-launching ramp. You will see more at the ones that have become state parks, but even there you will be pretty much restricted to trails on the dry perimeter and possibly a boardwalk over wet stretches.

One example of the latter is Jones Lake State Park, located on N.C. 242, 4 miles north of Elizabethtown. Included within its boundaries are two bays—Jones Lake and Salters Lake—a hiking trail, and extensive pocosin vegetation. In the 18th century this area was the focus of intensive exploration by botanists seeking to collect such oddities as Venus' Flytrap and Pitcher Plants, and nearby White Lake was known for many years as Bartram's Lake.

In South Carolina, Woods Bay State Park has an interpretive trail with 200 feet of raised boardwalk. Among the woody plants to be seen here are Sweet Pepperbush *(Clethra alnifolia),* Piedmont Azalea *(Rhododendron canescens),* Sweetleaf or Horse Sugar *(Symplocos tinctoria),* Red Buckeye *(Aesculus pavia),* and such vines as Climbing Hydrangea *(Decumaria barbara),* and the red-and-yellow Coral Honeysuckle *(Lonicera sempervirens).* This park is located 7 miles north of Turbeville off U.S. 301.

12

Exploring the Outer Banks

THE "OUTER BANKS" of North Carolina exemplify one of the most distinctive features of the Atlantic and Gulf coastlines: the barrier beaches. Thin slivers of sand less than a mile wide in most places, they appear on the map something like a drawn bowstring, pulled outward to a point near the middle. This extension, which puts Cape Hatteras 30 miles out from the mainland, is unique, however, as most barriers hug the coastline.

Barrier beaches may be peninsular, joined to the mainland at an extremity, or elongated islands where their continuity has been severed by the force of hurricane-driven seas. The importance of these tenuous strands in buffering the effect of violent storms upon the nearby interior can hardly be exaggerated.

Other barrier beaches occur both north and south of the Outer Banks and along much of the Gulf of Mexico as well, but they are not always consistent in form, changing in some places from a long strip to a chain of irregular islands separated from each other by inlets, and vanishing entirely in other stretches. The major interruption is in southern Florida, where a dense belt of mangrove swamps has taken over much of the shoreline.

The sand of most beaches is of mineral origin, usually quartz (which is found in nearly all types of rock) along with traces of other less common ones—each grain a tiny bit of the weathered mountains or piedmont brought down by rivers to the sea, and this holds true for the Outer Banks, which will serve to represent the middle Atlantic coastline in the following discussion. (In southern Florida things are very different, for there the beaches consist of the calcareous remains of organisms that lived in the warm waters that once covered the land.)

Most of the vegetation on these barrier beaches and islands is divided among three habitats: the dunes, the maritime forests, and the salt marshes.

SAND DUNES

Subjected to the full force of the ocean, salt spray, blistering heat, gale winds, and shifting sands, and lacking a stable substrate, the frontal dunes just above the beach itself support little vegetation. Most conspicuous on the primary dunes is Sea Oats *(Uniola paniculata),* native to the Atlantic and Gulf coasts of the southern United States, with flattened heads of ovate spikelets that turn dull gold in summer (this plant used to be dyed outrageous colors and sold by florists, but it now enjoys protected status). Its only rival as a sand binder is American Beach Grass, or Marram Grass *(Ammophila breviligulata),* which has a compact cylindrical inflorescence very different from that of Sea Oats. Both possess the ability to put out vertical rhizomes and produce new shoots even after repeated burial beneath the sand. The southernmost limit of the natural range of Beach Grass is Cape Hatteras, but it has been planted widely elsewhere.

A third species forms clumps, but is less effective and therefore survives better away from heavy sand buildup. It is known as Short Dune Grass *(Panicum amarum),* is conspicuous by the bluish green color of its leaves, and is distributed all the way to the Gulf of Mexico.

SEA ROCKET
Cakile edentula

One of the few forbs capable of eking out an existence on the fore dunes is the lavender-flowered Sea Rocket *(Cakile edentula).* Its answer to this salty, and therefore physiologically arid, environment is to store copious amounts of water in its thick, fleshy leaves and stems. Sea Rocket leaves are often used in salads, their pungency betraying membership in the Mustard family. Seaside Spurge *(Euphorbia polygonifolia)* defends itself against the wind by putting down numerous roots along its creeping stem, and reduces evaporation by having very small leaves. Sea Elder *(Iva imbricata)* is a shrub but is equipped with succulent foliage for the conservation of moisture in this desiccating environment.

The backs of the dunes fare a little better, and here as well as on the secondary dunes (provided they are lower and protected by the front dunes) we can find such plants as Glades Morning Glory *(Ipomoea sagittata),* at its northern limit here; the tall, thick-leaved Seaside Goldenrod *(Solidago sempervirens)* with its plume of yellow flowers; and Buttonweed *(Diodia teres),* a sprawling, pervasive plant with small but not unattractive pink flowers. Camphorweed *(Heterotheca subaxillaris)* sometimes rotates its petioles so that the leaf blades are presented to the sun edgewise—a means of reducing the loss of moisture by evaporation.

The principal grass on the back dunes is likely to be the valuable forage plant, Salt-meadow Hay *(Spartina patens),* although the long, needle-sharp barbed spines of Sandspur *(Cenchrus tribuloides)* are almost certain to claim one's attention.

MARITIME FORESTS

In sheltered areas behind the secondary dunes, woody plants manifest themselves in shrub thickets and maritime forests. Among the important members of these communities are Live Oaks *(Quercus virginiana),* often dwarfed and sheared by the accumulation of salt, which kills the buds exposed to the seaward side, Yaupon Holly *(Ilex vomitoria),* and Wax Myrtle, or Southern Bayberry *(Myrica cerifera).* The last grows to tree size, unlike the low shrubby habit of Northern Bayberry *(Myrica pensylvanica).*

Loblolly Pine *(Pinus taeda)* is the pioneer tree species and the principal pine on the Outer Banks, but it is not especially well equipped to endure salt spray and will grow only in areas that are well back from the ocean.

These shrubs and trees, the other species that eventually join them, and the more salt-sensitive plants that are able to grow under their protective canopy, are all at the mercy of the vegetation that has stabilized the intervening dunes. Should this disappear, the entire forest can be completely buried under the advancing sands of "walking dunes."

SALT MARSHES

Bordering the shallow sounds that separate the barrier islands from the mainland are the salt marshes. Here the winds are more moderate, and the peaty soil is more solid owing to the deposition of nutrient-rich sediments allowed by gentler tides and currents. One consequence of these relatively benign conditions is a far richer flora.

Salt-marsh Cordgrass *(Spartina alterniflora),* by far the dominant species, is the pioneer plant. It rapidly stabilizes the muck with its roots and forms a network for holding the silt and detritus washed back and forth by the tides and brought down by streams from landward. Also found here in dense stands on slightly higher land is Black Needlerush *(Juncus roemerianus),* a dark plant with cylindrical leaves terminating in short, extremely sharp spines, which actually were used by early settlers for sewing.

In the higher parts of the marshes there is Salt-meadow Hay *(Spartina patens),* which also occupies the back dunes, but here it expands into vast meadows, its fine wiry stems and leaves tangled and swirled by the winds into wavy patterns of bright green called "cowlicks." The short, pale green Salt Grass or Spike Grass *(Distichlis spicata)* is particularly salt tolerant, and thrives near poorly drained depressions called "pannes," where the salinity becomes concentrated as the accumulated seawater evaporates.

Of all the conspicuous flowering plants of the salt marshes, one of the prettiest is the pink Seashore Mallow *(Kosteletzkya virginica),* although Glades Morning Glory and Sea Ox-eye *(Borrichia frutescens)* add greatly to the scene. Less showy plants include Sea Lavender *(Limonium carolinianum)* and the shrubby Marsh Elder *(Iva frutescens)* and Groundsel Tree *(Baccharis halimifolia),* the latter conspicuous when in fruit because of its silky white pappus hairs.

THE OUTER BANKS

The best point from which to approach the North Carolina coastal dunes and associated wetlands is Whalebone Junction, at the northern entrance to the Cape Hatteras National Seashore. It is on Bodie Island (pronounced "body"), just east of Manteo at the juncture of U.S. 64 from the west, N.C. 12 from the south and U.S. 158 from the north, and therefore lies directly in the path of anyone traveling to or on the Outer Banks.

Jockey's Ridge

Driving north from Whalebone on U.S. 158 Bypass about 5 miles, you suddenly come upon enormous sand dunes looming above the highway on your left—the side *away from* the ocean. This is Jockey's Ridge, at 140 feet the highest dune on the East Coast, and a North Carolina State Park (the entrance is farther north).

From a recreational standpoint, the bare dunes are the sum total of the state park and can be enjoyed by sand skiiers, kite fliers, hang-gliding enthusiasts, and just plain dune climbers without hesitation, since there is no vegetation to be damaged. Around the base, however, are much more sensitive areas with excellent examples of gnarled Live Oaks and the ubiquitous Wax Myrtle struggling with the constantly shifting sands, and herbs like Slender-leaved Goldenrod *(Solidago tenuifolia),* Bitterweed *(Helenium amarum),* and Camphorweed *(Heterotheca subaxillaris).*

Of special interest is a small population of Woolly Beach Heather *(Hudsonia tomentosa),* common along the north Atlantic coast but at its southern limit here. This plant exhibits two adaptations for conserving water: small, appressed, almost scalelike leaves to reduce surface area, and a dense covering of insulating hairs. It also is efficient at trapping wind-blown sand despite its low, mat-forming growth habit.

Nags Head Woods

A couple of miles farther north, in Kill Devil Hills, is Nags Head Woods, an extensive maritime forest on the ridge and flanks of ancient dunes, maintained by The Nature Conservancy. The entrance is via Ocean Acres Drive to the west, which eventually becomes a dirt road and ends at the visitor center. The entire Nags Head Woods Preserve comprises 646 acres, but only a small portion is available to the general public and that only on certain days, so be sure to inquire in advance of your visit.

Two walks have been laid out, the shorter of which is known as the Center Trail and follows a quarter-mile loop with numbered stations corresponding to descriptions in a printed brochure. One of its nicest features, aside from some very attractive ponds, is a carpet of Partridgeberry *(Mitchella repens)* near the start of the trail (illustrated on the next page).

The Sweetgum Swamp Trail is considerably longer, and an excellent detailed guide is available (modest fees are charged for these printed materials). In addition to Loblolly Pine, and Sweet Gum *(Liquidambar styraciflua),* Black Gum *(Nyssa sylvatica),* and Red Maple *(Acer rubrum)* in the wet areas, these woods are notable for several species more commonly found in uplands, such as American Beech *(Fagus grandifolia),* Southern Red Oak *(Quercus falcata),* and Sweet Pignut Hickory *(Carya ovalis).* In the understory there are Flowering Dogwood *(Cornus florida),* Sassafras *(Sassafras albidum),* Muscadine Grape *(Vitis rotundifolia,* seen on p. 115), and Beauty-berry *(Callicarpa americana),* this last making a colorful show in autumn when the compact clusters of bright violet fruits ripen.

The first part of the trail consists of a sandy road where Spurred Butterfly Pea *(Centrosema virginianum)* and Wild Pink *(Silene caroliniana)* grow on the banks and several ferns, including Cinnamon, Royal, and Virginia Chain Fern (*Osmunda cinnamomea, O. regalis,* and *Woodwardia virginica*) can be seen in low-lying wet spots.

In the open sun at a power line crossing there is a profusion of Blue Curls *(Trichostema dichotomum)* and St. Andrew's Cross *(Hypericum hypericoides),* punctuated by that pretty white-flowered plant with nasty stinging hairs, *Cnidoscolus stimulosus.*

Shortly after reentering the woods there are two rather surprising species: Pink Lady's Slipper *(Cypripedium acaule),* which is usually found on higher ground, and the distinctively southern Spanish Moss *(Tillandsia usneoides).* These and the overlapping incidence of both the Southern and Northern Bayberries (*Myrica cerifera* and *M. pensylvanica*) remind us of the influence here of the warm Gulf Stream and the colder waters of the Labrador Current, both of which pass offshore.

Still more startling is an open sand flat in the midst of the forest, with pioneer dune plants like Short Dune Grass, Salt Marsh Aster *(Aster tenuifolius),* Coastal Chinquapin (*Castanea pumila* var. *ashei*), and the northern Woolly Beach Heather *(Hudsonia tomentosa).*

Last but not least are the freshwater ponds of Nags Head Woods, which are said to contain all four genera of those tiniest of flowering plants, the Duckweeds (*Spirodela, Lemna, Wolffia,* and *Wolffiella*) in addition to the strange-looking Water Violet, or Featherfoil *(Hottonia inflata).*

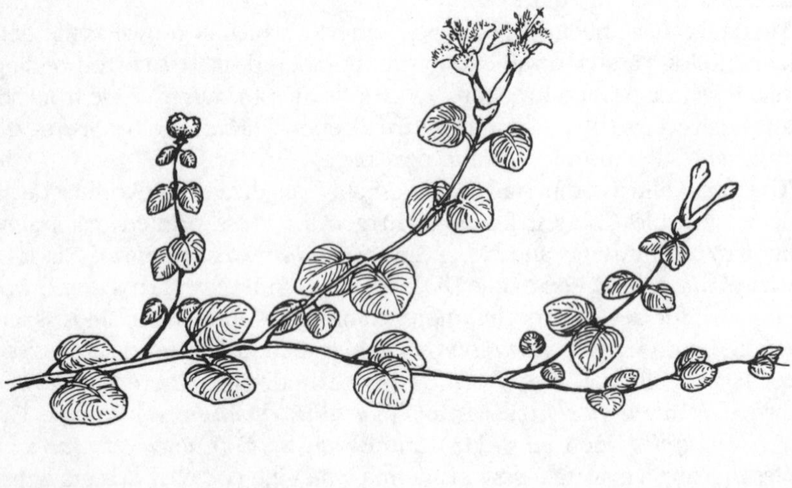

PARTRIDGEBERRY
Mitchella repens

MUSCADINE GRAPE
Vitis rotundifolia

Bodie Island Lighthouse

Going in the opposite direction, about 5.5 miles south of Whalebone on the west side of N.C. 12, is a road leading to Bodie Island Lighthouse. Just behind the lighthouse is the Bodie Island Pond Trail, a boardwalk leading to a fine example of a salt marsh.

The entrance is through a lush growth of Wax Myrtle, Groundsel Tree, and Poison Ivy *(Rhus radicans),* which is prolific but easily avoided, above a large bed of Marsh Fern *(Thelypteris palustris).* The boardwalk ends in a beautiful mixture of Cattail and Seashore Mallow, the bright pink flowers of the latter extending far across the marsh. Spike Grass grows next to the walk, with Salt-meadow Hay farther out.

Egrets, herons, and other wading birds are here in numbers; Wilson's phalaropes and avocets are to be seen during the September migration. One may often observe a nutria, a South American aquatic rodent intermediate in size between a muskrat and a beaver, feeding in or near shallow water.

There is a grove of Loblolly Pines within the traffic circle at the lighthouse, and a gated pedestrian road leads from the perimeter west to Pamlico Sound. The vegetation along this road includes Wild Black Cherry *(Prunus serotina),* Winged Sumac *(Rhus copallina),* Creeping Cucumber *(Melothria pendula,* seen on the next page), a delicate vine with diminutive yellow blossoms and half-inch fruits; Seaside Ground Cherry *(Physalis viscosa),* and Sea Ox-eye *(Borrichia frutescens).*

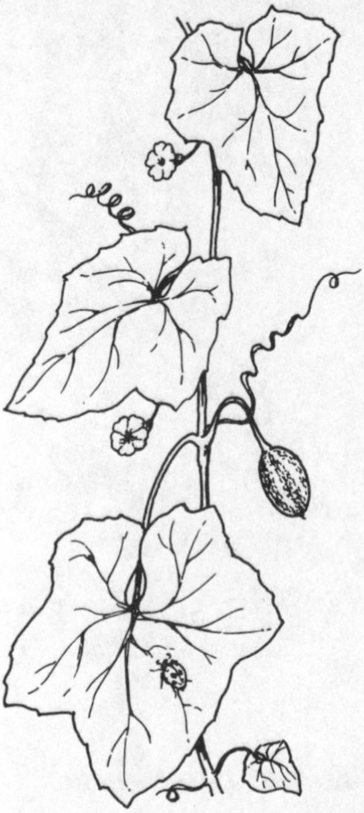

CREEPING CUCUMBER
Melothria pendula

On the shores of the sound are the rare Seaside Gerardia *(Gerardia maritima),* Marsh Pink *(Sabatia stellaris),* Salt Marsh Aster *(Aster tenuifolius),* Capeweed *(Lippia nodiflora),* and Glasswort *(Salicornia virginica).* Glassworts are peculiar plants with scalelike leaves and minute apetalous flowers, and are true halophytes, not only tolerating but actually demanding saline conditions. Consequently, they are often found in pure stands on mud flats where nothing else will grow. Their thick, fleshy, jointed stems and branches are so turgid with water that they are translucent, hence the common name, and not surprisingly the tender tips can be pickled.

Oregon Inlet

Three miles farther south on N.C. 12 is Oregon Inlet, where a hurricane in 1846 separated Bodie Island from Hatteras Island. The Fishery Center is a good place to become acquainted with such plants as Salt-marsh Bulrush *(Scirpus robustus)* and Marsh Sedge *(Fimbristylis spadicea).* Marsh Pennywort *(Hydrocotyle bonariensis)* is abundant here, and in many other locations as well, always creating a light green ground cover of scalloped round leaves.

Pea Island Refuge

The 3.5-mile-long Herbert C. Bonner Bridge over Oregon Inlet takes you to the Pea Island National Wildlife Refuge, which occupies the northern quarter of Hatteras Island. Although this is probably best known as a wintering ground for thousands of snow geese, it can also be rewarding for botanists.

About 7 miles south of Oregon Inlet there is a path encircling a large pond. A sign at the entrance reads North Pond Interpretive Trail, but there are no explanatory signs or brochures. Before entering the trail, you pass a flat area replete with Wild Bean *(Strophostyles helvola),* a trailing plant with pink-purple flowers that quickly turn to dull yellow, on erect stalks. Quite possibly this is the legume that gave its name to Pea Island (which, incidentally, was at one time separate from the rest of what is now called Hatteras Island).

Many of the species typically associated with brackish marshes in this area are found farther along the trail, but some others noted were the white-flowered Buttonweed *(Diodia virginiana),* a second species of *Lippia (L. lanceolata),* known as Fog-fruit; and on a dry sandy bank one of the Dayflowers, *Commelina erecta.*

Cape Hatteras

Nearly everyone will want to visit Cape Hatteras, with its imposing 208-foot brick lighthouse (the tallest in the United States) and its fine wide beach looking out toward Diamond Shoals, the treacherous "graveyard of the Atlantic," the bane of mariners for centuries. Two left turns from N.C. 12 near Buxton at signs for the lighthouse bring you to a parking area near the base of the structure.

Behind the rows of Sea Oats near the sea there is a rich assortment of back dune plants, Gaillardia *(Gaillardia pulchella),* Seaside Evening Primrose *(Oenothera humifusa),* and Camphorweed being among the most conspicuous. Prickly Pear Cactus *(Opuntia drummondii)* is here, and its easily detached, long-spined pads frequently hitch a ride on the sneakers of the unwary. These fleshy segments are modified stems; the leaves are minute and are deciduous. Vines are everywhere, including Pepper-vine *(Ampelopsis arborea),* Milkweed-vine *(Cynanchum palustre),* Milk-pea *(Galactia macreei),* and Glades Morning Glory.

A startling demonstration of the extent to which these beaches are subject to erosion can be seen at Cape Hatteras. When the lighthouse was built in the 1860s it was 1,000 feet from the sea; it is now less than one-tenth of that distance, and will be saved only if the desperate measures recently undertaken eventually win the battle with the sea.

Buxton Woods

Following the directions to Cape Hatteras Lighthouse but turning right instead of left at the second sign brings you to Buxton Woods, a part of the 30,000-acre Cape Hatteras National Seashore. Here a three-quarter-mile self-guiding nature trail loops through another maritime forest, where there is ample evidence of the difficult time that trees have coping with wind, salt spray, and a substrate that is physically unstable and nutritionally poor.

Live Oaks are here along with Loblolly Pine and Red Cedar. Many understory trees and shrubs can be seen, such as Flowering Dogwood, Ironwood *(Carpinus caroliniana),* Yaupon and American Hollies (*Ilex vomitoria* and *I. opaca*), Devil's Walking-stick *(Aralia spinosa),* Beauty-berry, and Wax Myrtle, as well as vines like Greenbrier *(Smilax bona-nox)* and Muscadine Grape.

Wet areas produce a few Bald Cypress trees *(Taxodium distichum),* Marsh Fern *(Thelypteris palustris),* and Netted Chain Fern *(Woodwardia areolata).* Dwarf Palmetto *(Sabal minor),* the northernmost of our East Coast palms, is near the upper extremity of its range here in Dare County.

Herbaceous wildflowers are few, some exceptions being Horsemint *(Monarda punctata)* in sunny openings and Elephant's-foot *(Elephantopus tomentosus)* along the trail edges. Three Birds Orchid *(Triphora trianthophora)* is unusual and a welcome find anywhere, but here in Buxton Woods it is a disjunct, at least 300 miles from its next nearest occurrence in the mountains to the west.

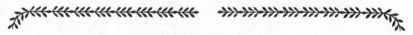

13

A Touch of the Tropics

TO VISITING BOTANISTS, Florida south of Lake Okeechobee is nothing if not perplexing. Suddenly to be confronted with palm trees, epiphytic orchids, and most frustrating of all, an apparently endless array of unfamiliar woody plants with unlobed, undivided, untoothed, and altogether undistinguished leaves—yet obviously all different—convinces them that they are in an entirely new world and that, since it is warm, it must be the tropics. This conviction is reinforced if they know something of the flora of the West Indies, for they will immediately recognize identical species here. Furthermore, the region obviously has the two-season climate typical of the tropics.

One look at a map, however, will tell them that this cannot be, for by definition the tropics begin at the Tropic of Cancer and this lies a hundred miles to the south. Also, it is well-known that even southern Florida is subject to cold snaps—definitely a nontropical attribute. And what about the dozens of species from the north that *do* seem content to make this their home?

The truth is that, for a complexity of reasons, southern Florida is able to provide satisfactory habitats for many plants of both the tropical *and* the temperate zones, and is therefore one of the most diverse botanical areas in the world. To appreciate the extent of this variety, just consider that this piece of land no part of which rises more than 25 feet above sea level contains a coastal mangrove belt, sawgrass prairies, tropical hardwood hammocks, cypress swamps, pinelands, and gleaming sandy seashores.

There is one specific area that, although it constitutes but a small part of southern Florida, contains examples of all six: Everglades National Park. Since most of our mangrove swamps and practically all sawgrass prairies are within its boundaries, these two biomes will be discussed in the chapter devoted to the park.

No examination of southern Florida's natural vegetation would be complete without some attention to a couple of factors that have exerted a substantial influence upon it. One is the intensive use of the land for the growing of crops, and the other the introduction of exotic and, as it turns out, highly undesirable plant species.

FRUIT AND VEGETABLE FARMING

Travelers to the Everglades National Park by way of Homestead, or to the Keys, find themselves in the midst of the southernmost farming region in the continental United States as they pass through Broward, and especially Dade, counties. Called "Redlands" because of the color of the soil, this expanse of perfectly flat land consists chiefly of limestone rockland and marl, which are poor in organic content and must be heavily fertilized and irrigated.

Acreage devoted to vegetables is the most extensive, with beans, tomatoes, and potatoes in the lead. In terms of cash value, one of the major crops is Boniato. This is the tropical American Sweet Potato, and is a member of the Morning Glory family with the scientific name of *Ipomoea batatas* (not to be confused with the unrelated Yams, which are in the genus *Dioscorea*). Many fields are opened up for customers to pick their own vegetables during the winter months.

As might be expected, Dade county is foremost in the production of tropical fruits, the leaders being Avocado (*Persea americana,* a relative of Red Bay), Lime *(Citrus aurantifolia),* Mango *(Mangifera indica),* Papaya *(Carica papaya),* and Plantain (*Musa* hybrids), in that order. Utilizing much less acreage but very important economically is the raising of ornamental nursery stock.

There is an excellent resource for learning more about southern Florida agriculture and the exotic plants in which it specializes in Homestead. It is the Preston B. Bird and Mary Heinlein Fruit and Spice Park, a living museum where you can become acquainted with 500 different fruits, spices, nuts, and economic plants, and may even sample them from the ground. A special section has been set aside for toxic plants and is equally instructive. The facility is operated by the Metro-Dade County Park & Recreation Department, and guided tours are conducted, but inquiries as to dates and times should be made in advance. A guide book containing a wealth of information regarding both edible and harmful plants is available. The park is located on Coconut Palm Drive, 1 mile west of Route 997 (which is also known here as Krome Avenue).

It is also possible to take a guided bus tour of south Florida farms from the government-owned State Farmers Market, which is situated on Krome Avenue in Florida City. These tours are restricted to certain days of the week from December through March.

ALIEN PLANT INTRODUCTIONS

The section of the Tamiami Trail (U.S. 41) between Florida's Turnpike and Naples is one of those straight, flat highways that seem to demand high-speed driving (another being the toll road running parallel to it farther north known as Alligator Alley, or S.R. 84). Yet any motorist who is at all concerned with the passing landscape will not fail to notice the existence of several invasive plants in vast numbers.

Dense rows of feathery, dark green Australian Pines form admittedly attractive screens along one side of the road. These are not pines at all but species of *Casuarina* (*C. equisetifolia* is the one most commonly seen in Florida), although one could be excused for applying the name to them on the basis of the superficial appearance of their "foliage" and the presence of small woody cones. What appear to be long needles are in fact jointed twigs, which remain green and perform photosynthesis; the leaves are reduced to minute scales whorled at the nodes and visible under a hand lens.

These trees were imported into Florida from the southwest Pacific late in the 19th century, apparently for their ornamental value. Their extremely rapid growth suggested a utility for stabilizing soil in the wake of human disturbance of the land. Eventually, however, the disadvantages became apparent. The trees' roots extend down to the tide line and render beach areas unsuitable as nesting sites for rare crocodiles and sea turtles, but the root systems are not deep enough to keep the trees from toppling under the onslaught of hurricane-force

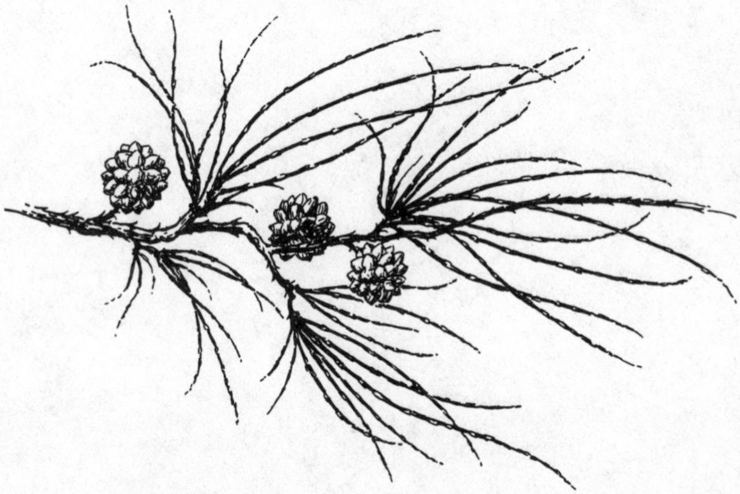

AUSTRALIAN PINE
Casuarina equisetifolia

winds. It has also been discovered that the fallen branches exude a toxin which inhibits the growth of other plants in the area.

Another exotic forms mats in the water of the Tamiami Canal, which runs alongside the highway. If Australian Pine is considered handsome, the Water Hyacinth *(Eichhornia crassipes)* is downright beautiful. Its smooth round leaves have stalks that are swollen at the base, appearing inflated but actually filled with spongy tissue. The inflorescence is a crowded spike of large irregular flowers, lavender with a yellow blaze on the upper lobe (larger than, but reminiscent of, those of the related Pickerel Weed).

It is not at all surprising that these pretty blossoms were the cause of the plant's having been introduced into this country. The story is that quantities were brought from Brazil to the 1884 Cotton States Exposition in New Orleans and given as souvenirs to visitors at one of the pavilions, and that a woman innocently threw hers into the St. Johns River.

What no one knew was that Water Hyacinth is one of the fastest-growing plants on earth and that, in the absence of natural factors that controlled its spread in South America, it would form continuous mats choking waterways, impeding navigation, causing flooding, and blocking sunlight from reaching

WATER HYACINTH
Eichhornia crassipes

other aquatic plants. In the United States it now covers an estimated two million acres. Recently it has been found that Water Hyacinth is extremely efficient in absorbing waterborne pollutants, and this has raised the hope that it might someday be useful in the treatment of wastewater.

A third alien species is *Melaleuca quinquenervia*, the Cajeput Tree. Another native of Australia, it was introduced into Florida for landscaping in the early 1900s and is now found in many parts of the state. A good example will be seen along the Tamiami Trail near Monroe Station, where it is becoming a threat to the Big Cypress Preserve.

This is a case where a plant's qualities were known in advance but the consequences not fully realized. It grows rapidly, absorbs vast quantities of water, and is resistant to fire—all thought at first to be desirable properties. It soon became evident, however, that the trees' production of seeds by the millions meant solid thickets in no time at all, that their appetite for moisture left the soil so dehydrated as to be uninhabitable by any native plants, and that attempts to destroy trees by burning not only failed but actually stimulated the reproductive process. With its white, spongy bark and cylindrical spikes of creamy "bottlebrush" flowers, *Melaleuca* is instantly recognizable.

Yet another beauty turned beast is Brazilian Pepper *(Schinus terebinthifolius),* a fast-growing woody plant that in less than 40 years has taken over large areas, eliminating nearly all other vegetation in its path. Its red berries are undeniably attractive, and their appeal has not been lost on seed-eating birds that have contributed greatly to its distribution. What apparently was either unknown or ignored at first is that it has toxic properties similar to those of Poison Ivy.

All is not discouraging along this road, however. On the south side, about 14 miles east of the junction with S.R. 29, there is a refreshingly attractive picnic area. Here a semicircular boardwalk leads through a miniature cypress swamp with Fire Flags and a good many other flowering plants.

Wading birds are active in the canal, but it is difficult to see them while driving. There is no such problem with the belted kingfishers, though; it seems as if the nation's entire population of these jaunty birds must be distributed along the Tamiami Trail's overhead wires.

14

Everglades National Park

THE WORD "EVERGLADES" is said to have entered our language by way of an error in transcribing "river glades" from an 18th century map. It has come to mean a region of low land, usually flooded with water and covered with grasslike vegetation, and to be sure, such wet prairies account for much of the region we find marked "The Everglades" on maps of southern Florida. (As we have noted, there are also hardwood hammocks, pinelands, and cypress swamps here. However, they owe their existence to subtle but definite variations in elevation, and as a consequence they are found in other parts of the South as well. For this reason, each is accorded a chapter of its own.)

Everglades National Park takes up only a third (albeit 1.4 million acres) of the Everglades, and much of that is inaccessible for most practical purposes. Yet the portions of it that can be visited comfortably represent a remarkable cross section of the whole and are an excellent showcase for the subtropical vegetation that in the United States is found only here.

The drive through the park—and there is only one road, 38 miles long—starts at the Main Visitor Center southwest of Homestead and terminates at Flamingo. (Here it should be pointed out that the highway shown on some maps as S.R. 27 has been redesignated. North of Florida City it is now Route 997, and between Florida City and the park entrance it is Route 9336.)

To really see the Everglades, a minimum requirement is that you stop frequently along the park road. Certainly you will want to do this where there is access to trails, but you should not overlook the fact that the roadsides themselves exhibit a variety of plant life much of which can be missed when you are cruising along even at moderate speed.

Where the moist prairies come up to the road shoulders there are drifts of *Sabatia grandiflora,* one of the most beautiful of the Marsh Pinks. To describe these flowers as having a yellow, crimson-bordered blaze set amid bright rose-pink petals is to make them sound excessively gaudy, but nature has brought it off very tastefully. To harmonize with their principal color there are species of Wild Petunia *(Ruellia)* and Blue-eyed Grass *(Sisyrinchium),* Bluehearts *(Buchnera americana),* and the more delicate Salt-and-Pepper *(Melanthera hastata)* and Water Pimpernel *(Samolus ebracteatus),* with the long-pointed bracts of White-top Sedge *(Dichromena colorata)* providing graceful accents.

Visitors from the North, accustomed to watching the last of the year's flowers succumb to frosts in late autumn, are often startled to find goldenrods and thistles brightening the southern landscape in February. What looks like a narrow-leaved, flat-topped Goldenrod in the glades, however, is likely to be a closely related composite known as Yellowtop, or *Flaveria linearis,* a plant also common southward through the Keys and beyond to the West Indies. The common Thistle has large heads the color of strawberry ice cream, and it is difficult to believe this is the same Horrid Thistle *(Cirsium horridulum)* that in much of its range, which covers two coasts from Maine to Texas, is dull purple or yellowish.

Although winter is traditionally the season when many try to escape from ice and snow to the balmy climate of the "Sunshine State," there is a particular reason for traveling to the Everglades at that time of year: It is the *dry* season. This is the period when wildlife of all kinds is forced to concentrate in the comparatively wet places and therefore can be observed more easily and in greater numbers. It is also the time when the insect population is at its nadir. This is an advantage that can be fully relished by anyone who has been here in the summer, when, it has been said without much exaggeration, it is possible to wave a pint jar in the air and come up with a quart of mosquitoes.

SAWGRASS PRAIRIES

The sawgrass glades (or prairies, in the local sense of that word) extend southward from Lake Okeechobee toward Florida Bay. The Seminole Indians have called the region "Pa-hay-okee," meaning "river of grass," an apt description since for most of the year it is flooded by a shallow, very slowly moving sheet of fresh water draining from the 700-square-mile Lake Okeechobee, and its vegetation appears predominantly grassy.

Winter and spring are normally dry, and there are cyclic droughts, but these natural tribulations have been intensified during the past several decades by human interference. Reclamation, irrigation, and flood control measures have drastically altered the drainage patterns and have even caused the disastrous intrusion of salt water from the Gulf. During dry periods there are frequent fires in the arid prairies, and water is then available to wildlife only in shallow troughs called sloughs (pronounced "slews") or in gator holes that are excavated by alligators where the water table is near the surface. In such places insects, fish, birds, mammals—just about every kind of animal from one end of the food chain to the other, ending with the omnivorous alligator—congregate in a frenzy of feeding.

The principal component of these prairies, and the one that gives the name Sawgrass to this plant community, is not really a grass but a sedge, *Cladium jamaicensis.* If "grass" is not appropriate, the "saw" part is, for the two edges and the back of the midrib have rows of extremely fine, sharp teeth—about 30 to the inch—that cut like a razor and go a long way toward discouraging progress on foot.

Anhinga Trail

This is an elevated walk over Taylor Slough, probably the best place in which to observe at close range not only the birds of the Everglades but such strange creatures as soft-shell turtles, the primitive garfish, and of course alligators.

Edging the open water areas here are masses of Spatterdock, Arrowhead, and Pickerel Weed *(Pontederia cordata),* and wetland shrubs ranging from the strictly southern Pond Apple *(Annona glabra)* to Buttonbush, whose range extends all the way to maritime Canada. As one would expect, there are anhingas: some diving, some preening, some hanging their wings out to dry. The magnificently colored purple gallinules weave their plodding way in and out of the rushes past stalking little blue herons and probing white ibis.

For most of its distance, the boardwalk passes through Sawgrass flats, at midpoint going through a Willow head where there are a few Coco Plums *(Chrysobalanus icaco).*

SWAMP LILY
Crinum americanum

Pa-hay-okee Overlook

This elevated platform, reached by a boardwalk, provides one of the best views of the Sawgrass glades in the park. It looks out over the Shark River Slough, the main, and critical, source of water for the Everglades.

The curving white tepals of Swamp Lily, or String Lily *(Crinum americanum),* stand out conspicuously amid the Saw Grass and can be seen over long distances. Nearby there are white umbelliferous plants called Water Dropwort *(Oxypolis filiformis),* with curiously jointed hollow leafstalks without blades.

In the short grass around the parking area you may see a low-growing terrestrial orchid with a compact spike of white or yellowish flowers. This is *Zeuxine strateumatica,* an Asian introduction that has become naturalized throughout Florida and is known familiarly as Lawn Orchid because of its predilection for grassy or disturbed sites. It blooms in the winter here, and farther north is one of the earliest of spring flowers.

MANGROVE SHORES

Fringing the shores of the lower part of the Florida peninsula and the Keys are the great mangrove swamps. The name "mangrove" derives from three different species of evergreen trees that together form the major part of these dense, impenetrable thickets: Red Mangrove *(Rhizophora mangle),* Black Mangrove *(Avicennia nitidia),* and White Mangrove *(Laguncularia racemosa).* Their presence here is due to their need for a climate that is free of frequent frosts, a degree of tolerance for salt water, and the ability to grow in sandy or muddy shallows.

Red Mangrove (see the next page) is the pioneer species and occupies the outer, or seaward, rim of vegetation. As we shall see, it also is the plant responsible for progressively extending the shoreline farther and farther out into the water by trapping debris among its roots.

A distinctive feature, which makes it easy to recognize, is its system of arched stiltlike roots rising out of the water; these not only furnish a stable support for the plant in its watery habitat but also contain tiny openings on their exposed surfaces through which essential oxygen is absorbed from the air. These roots also provide a home for coon oysters, and their encrusting shells are revealed when the tide is out. These constitute one of the many delectable foods for the raccoons inhabiting the area.

In its reproductive process, Red Mangrove exhibits an amazing adaptation of a plant to its environment. The yellow flowers produce inch-long fruits that germinate while still attached and send out slender cigar-shaped radicles that may attain a length of 12 inches before dropping off. The radicle gradually becomes heavier at its pointed bottom end, and although it may float horizontally for some time it ultimately will embed itself upright in the mud, whereupon it

RED MANGROVE
Rhizophora mangle

will immediately emit leafy shoots from the emersed upper end. As the seedling grows into maturity, sediment and decomposing organic material begin to collect in the tangle of its interlacing prop roots, and as these deposits build up and spread out, the water around the plant gradually becomes more shallow. This is just what is needed for *its* seedlings to implant themselves in the unoccupied space to seaward, and thus another rank of mangroves becomes established. Others will float away on the tide, drifting for many months and many miles before starting a new colony on a distant shoal.

Black Mangrove also forms dense, often impenetrable thickets, but since it is less critical as to habitat it tends to range farther north. Whereas prop roots are the field mark of Red Mangroves, Black Mangroves are surrounded by multitudes of erect fingerlike pneumatophores, or breathing roots, sticking out of the mud; these perform the respiratory function attributed by some authorities, at least, to the more familiar "knees" of Bald Cypresses.

Black Mangrove belongs to the Verbena family and has small white flowers in terminal clusters. Its fruit is a compressed ovoid capsule and, like that of Red Mangrove, is ready to take root upon falling into the mud.

The third species, White Mangrove, occupies higher, more solid ground toward the interior of the swamp. It has greenish flowers, smaller than the others, arranged in spikes.

Buttonwood *(Conocarpus erectus)* is in the same family as White Mangrove and shares the same habitat. The generic name alludes to the small fruits, which are reddish and conelike. There is only one species, but an attractive form with silvery foliage var. *sericea* occurs in southern Florida and the Caribbean.

West Lake

Paurotis Pond, 5 miles beyond Mahogany Hammock, marks a major transition—from the vast sawgrass everglades to the world's largest mangrove forest. The lakeside is a quiet, secluded place where alligators doze along the banks and the charming little marsh rabbits graze, both almost oblivious to human visitors.

It is West Lake, however, that is one of the treasures of the park's mangrove community. A half-mile interpretive trail threads its way through all four kinds of "mangroves"—Red, Black, and White as well as Buttonwood—and one could hardly wish for a better classroom in which to study them.

Among the other plants to be seen along the walk, one that is especially abundant is Matrimony Vine *(Lycium carolinianum)*. Its extremely long, arching stems bear lavender blossoms, but these are almost completely obscured by the multitude of short leaves, and give way to brilliant red fruits that have earned it the alternate name of Christmas Berry.

DWARF CYPRESSES

Just west of Pa-hay-okee Overlook, there is a short boardwalk leading to an exhibit that describes an eroded outcropping of Miami oolite known as "pinnacle rock." To an ant, these grotesque formations might resemble the sandstone structures we see in places like Arizona or Utah, but in reality they are just inches high. In fact, the outcropping itself, elevated enough above the surrounding prairie to have earned the name "Rock Reef," is only 3 feet above sea level.

Beginning near here and continuing for quite a distance there is an area of scattered dwarf Bald Cypress trees. These are the typical species, *Taxodium distichum,* but it is thought that their size is kept down by the paucity of nutrients in the thin layer of soil overlying the marl.

FLAMINGO

The road ends at Flamingo, where there is another visitor center as well as camping facilities, and a concessioner-operated complex that includes an inn, restaurant, and marina, and provides boat rentals and tours.

As is to be expected in a developed area, a great many trees have been planted for their aesthetic value or botanical interest, or both. There are graceful Coconut Palms *(Cocos nucifera),* Strangler Figs *(Ficus aurea)* and Cabbage Palms *(Sabal palmetto)* in various stages of combat, rows of West Indies Mahogany *(Swietenia mahagoni),* and several clumps of Everglades Palms *(Acoelorrhaphe wrightii).*

Near the campground is Eco Pond, a water treatment facility but an attractive natural area nevertheless. Southern Cattails *(Typha domingensis),* a narrow-leaved species with a space between the staminate and pistillate portions of the flower spike, form a dense screen in the shallows. Back a little distance from the water the Buttonwoods are overgrown with the high-climbing Moon Flower *(Ipomoea alba).* Another vine, Climbing Hempweed *(Mikania cordifolia),* makes sprawling mats at somewhat lower levels. Blodgett's Nightshade *(Solanum donianum)* flourishes in the drier areas. This is a gray-green shrub with panicles of white flowers and scarlet berries.

SOUTHERN CATTAIL
Typha domingensis

Eco Pond is one of those places where you should use considerable caution when approaching the water's edge, out of respect for the numerous alligators that like to bask on the weedy banks.

Another area to explore near Flamingo is the road along Buttonwood Canal, which can be driven as far as Bear Lake Road. An interesting plant that is plentiful here is Nicker-bean *(Caesalpinia crista),* a viny shrub with bipinnate foliage and covered with hooked thorns. The large brown pods, which are also covered with prickles, hold two smooth gray seeds. These are often found on seashores where they have been washed up by the tide, and for this reason are called sea beans.

Near the boat dock on Bear Lake Road it is possible to see the notorious Manchineel *(Hippomane mancinella),* a deceptively handsome tree whose milky sap causes the skin to blister on contact. The Carib Indians used it as an arrow poison, and this is thought to have been the cause of Ponce de Leon's death, although many other tales of recalcitrant sailors and soldiers having been stripped and chained to the trees as punishment by the conquistadors may be apocryphal. Nevertheless, it is extremely toxic and should be avoided. Above all, the temptation to sample the fruits—which resemble crabapples—must be resisted, as they bring about complete destruction of the mucous membranes.

Farther along Bear Lake Road there are some Cinnamon Bark trees *(Canella winterana).* This is an evergreen, rare in the Everglades but common in the West Indies (the famous Caneel Bay in the U.S. Virgin Islands is named for it). Its bright yellow inner bark is pleasantly aromatic and is used as "wild cinnamon" in medicine and as a condiment, although most of the cinnamon we employ as a spice comes from Asiatic species of *Cinnamomum.*

Several trails up to a dozen miles in length are located in this end of the Park, and conducted tram tours operated out of Flamingo are an excellent way to experience some of them.

SHARK VALLEY

There is a second gateway to the park at Shark Valley. (Actually there is a third, at Everglades City on the Gulf coast, but it is intended for those who wish to explore the watery maze of the Ten Thousand Islands either by renting a boat or by taking a guided boat tour.)

The Shark Valley entrance is located on the Tamiami Trail, U.S. 41, about 18 miles west of Route 997 (remember, this was formerly S.R. 27) near the entrance to the Miccosukee Indian Reservation. From there, access into the park is on a 15-mile loop road that is limited to hikers, bicyclists, and trams. The tram tours (which are conducted by ranger-naturalists) are by far the most popular mode of travel on the loop road, and reservations must be made well ahead of time.

As an alternative—and a good one for anybody interested in the flora—there is the Bobcat Trail, which utilizes some segments of the loop road but also has its own foot trails through hammocks and across marl prairies. Conspicuous in the latter near the start of the trail are the lacy plants of Water Dropwort, accompanied by Lawn Orchids and Bluehearts along the road edges. The path soon turns and enters a bay head, passing through Coco Plum, Red Bay, Wax Myrtle, Coastal Plain Willow *(Salix caroliniana),* and Dahoon Holly *(Ilex cassine)* on a boardwalk, then emerges into an open, wet area with Arrow Arum *(Peltandra virginica)* and Pipewort growing amidst the Sawgrass, and ultimately joins the loop road.

Here a narrow canal runs alongside, harboring Pond Apples and Willows, the water in some places nearly obliterated by Spatterdock, Pickerel Weed, and the gigantic leaves of Fire Flag *(Thalia geniculata).* The banks are a tangle of dull pink Climbing Hempweed accented by the showy yellow flowers of Primrose Willow *(Ludwigia peruviana).*

On one occasion two gorgeous purple gallinules put on an amusing show here as they clambered drunkenly through one willow after another to feed on the staminate catkins, doing their best to climb with those long, splayed, and entirely unsuitable golden toes.

On the opposite side, the progression between road and thicket runs from White-top Sedge, Water Pimpernel, and pink Thistles through a rank growth of 6-foot high Bracken, then to trees like Myrsine *(M. guianensis)* draped with

SPANISH NEEDLES
Bidens alba var. *radiata*

Woodbine *(Parthenocissus quinquefolia)* and Poison Ivy *(Rhus radicans).* Occasionally a smooth, almost white trunk gives away a Sweet Bay *(Magnolia virginiana).* Spanish Needles, which in this part of the South means the white-rayed *Bidens alba* var. *radiata,* is here, of course, as it is in so many other sunny spots. Everywhere in Florida this flower seems to be a favorite of the zebra butterfly, although its larvae feed on leaves of the purple Passion Flower *(Passiflora incarnata).* This butterfly is our sole representative of the tropical group known as heliconians, and its elongated, boldly striped wings immediately mark it as being different from all our other species.

There are numerous gator holes near the road, and it is not at all difficult to find one or more of the armor-plated saurians close by.

At a sign reading Otter Cave the trail turns left on a short spur into a hammock. Here there is an open area with a floor of Miami oolite pockmarked by small solution holes. Understandably, this is devoid of vegetation except where an occasional plant has managed to get started in one of these little depressions, but around the perimeter where much larger and deeper deposits of humus have accumulated there is a grove of Gumbo Limbo trees *(Bursera simaruba)* and Pigeon Plums *(Coccoloba diversifolia)* mixed with Strangler Figs.

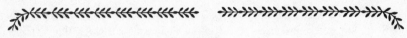

15

Tropical Hammocks

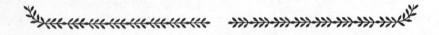

THE TERM "HAMMOCK" as used in Florida simply means a forest composed mostly of tropical hardwood trees and shrubs and growing on an elevated area. Because they contrast so markedly with the sea of grassy vegetation that surrounds them, hammocks are sometimes referred to as "tree islands" or "keys."

Since most of the plants on these hammocks are of West Indian origin, many cannot be viewed anywhere else in the United States, and consequently they hold a great deal of interest. One valuable species that was abundant here originally but has been extensively lumbered is West Indies Mahogany *(Swietenia mahagoni).* The fruiting capsule of this tree is unusual in that it splits open from the base upward.

A more common sight is the native Gumbo Limbo *(Bursera simaruba),* a member of the same family as the trees whose resins produced frankincense and myrrh. It is the most massive tree in southern Florida, and the most easily recognized by reason of its exfoliating older bark, which is thin, skinlike, and attains a lustrous cinnamon color. Floridians call it the "tourist tree" because, as they say, it turns red and peels.

Another interesting tree is Jamaica Dogwood *(Piscidia piscipula),* a legume with pinkish papilionaceous blossoms and papery-winged fruits. It is also known as Florida Fishpoison-tree from the Carib Indian practice of stunning fish with a preparation made from its bark. This is one of many tropical trees that produce their flowers on bare branches after the periodic shedding of their leaves.

The hammock tree with the most unusual habit would have to be the Strangler Fig *(Ficus aurea).* Its life cycle begins when a sticky undigested seed voided by a bird lands on the limb of another tree. Upon germination, a succession of aerial roots descend to the ground surrounding the host's trunk, while branches grow upward to form a leafy crown. Over the years the aerial roots expand and coalesce, strangling the host, but meanwhile the fig tree has managed to establish an independent existence.

Probably the most abundant tree is Poisonwood *(Metopium toxiferum),* related to Poison Ivy and at least as virulent. Fortunately, though, the midrib and edges of its glossy green leaves are bright yellow, and once this is learned it becomes easy to steer clear of it.

Because of the deep shade under the tree canopy, vines that climb to reach the sunlight are numerous in all tropical forests, and Florida's hammocks have their quota, one of the most handsome being Moon Flower *(Ipomoea alba),* shown on the next page. As the scientific name would indicate, it is a white morning glory, but there are other characteristics to which this gives no clue. Its size, for example: The flowers measure nearly 6 inches in length. For another thing, they open at night, and so rapidly that you can actually observe the unfurling of their spirally twisted buds. Still another feature is that the plant thrives in burned-over areas, growing prodigiously in the wake of hammock fires.

The vegetation of the hammocks is not restricted entirely to tropical species, and a number of trees from the temperate zone have extended their ranges into this subtropical habitat, among them Sugarberry *(Celtis laevigata)* and Red Mulberry *(Morus rubra).*

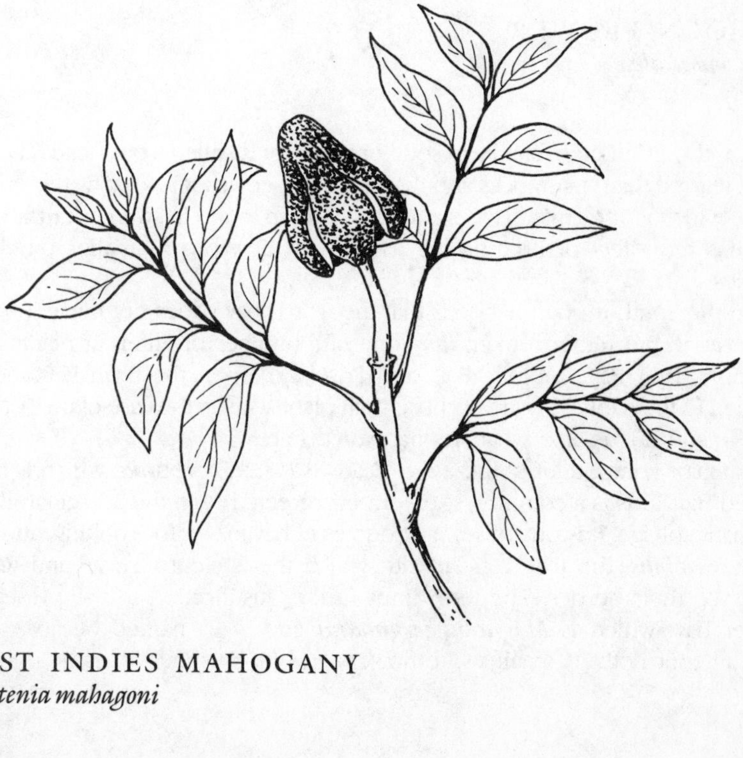

WEST INDIES MAHOGANY
Swietenia mahagoni

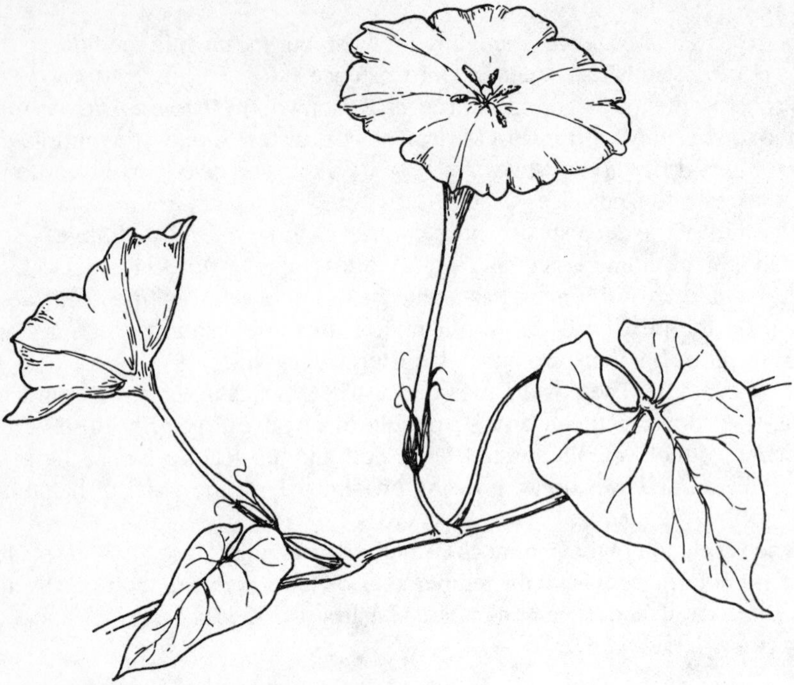

MOON FLOWER
Ipomoea alba

Another islandlike feature in southern Florida is called a bay head. These are less elevated than hammocks and therefore present a somewhat wetter environment, which is attractive to trees and shrubs often found in swamps farther north but has the effect of holding the assortment down to a smaller number of species.

In the South it often happens that the word "bay" does not mean a body of water at all but rather a plant (any one of a number of different plants, as for example, Bull Bay, Loblolly Bay, and Rosebay). Then the term is sometimes applied to the habitat of one or more such plants—as in the case of the "Carolina bays"—and this is also what has happened here.

A major component of these bay heads is *Persea borbonia,* which is known as Red Bay. This is a common southern evergreen tree in the Laurel family with aromatic foliage. It is one of several sources of bay leaves for cooking; another is *Laurus nobilis,* the true Laurel, with which the ancient Greeks and Romans crowned their heroes. The term finds further justification in the presence of Sweet Bay, which is *Magnolia virginiana* and is so named because of the fragrance of both its small white flowers and its crushed leaves.

In other places a slight accumulation of soil will provide a footing for the Coastal Plain Willow *(Salix caroliniana),* and the term for such congregations of these trees is, predictably, "willow head."

Where shallow ponds form, Bald or Pond Cypress trees provide yet another variation in the otherwise monotonous sawgrass landscape. Groves of these cypress trees are called cypress domes because of the rounded contours of their tops.

ROYAL PALM HAMMOCK

Upon entering Everglades National Park, the first opportunity to see other than roadside vegetation comes at Royal Palm Hammock, appropriately named for the splendid Florida Royal Palms *(Roystonea elata)* to be seen there. These stately trees have distinctive gray trunks looking very much like concrete, as do several other species of Royal Palm from the West Indies that have been introduced throughout southern Florida.

At this stop there are two trails, one of which is the Anhinga Trail, described in Chapter 14. The other is the Gumbo Limbo Trail, where rich hammock vegetation can be studied, and as at most such points in the park some of the plants are labeled. Along with the orange-brown trees that have given this trail its name, there are Strangler Figs, Wingleaf Soapberry, Snowberry, Wild Coffee, and Pigeon Plum *(Coccoloba diversifolia).* The last is a lesser-known relative of the Sea Grape, and its dark red, astringent fruits are favored by white-crowned pigeons, which accounts for the common name.

One large tree leaning over the trail is replete with air plants (*Tillandsia* spp.) as well as a patch of *Peperomia magnoliaefolia.* Epiphytic orchids were also common here at one time, but any that remain will be high in the trees, well out of reach of would-be collectors.

At one point there is a good example of a solution hole, where acids produced by the decomposition of organic material have dissolved the limestone and eventually created a basin that now serves as a water hole for wildlife. This also makes it plain that the water table is only a couple of feet below the surface of the rock. When they are mere pockets barely large enough to hold a handful of humus, these solution holes act as nurseries for many kinds of plants, especially ferns. The most abundant one here is Wild Boston Fern, but Maiden Fern *(Thelypteris normalis)* is also in evidence.

Along the outer edge of the hammock there are two plants that southern Florida shares with the tropical Americas. One is a prolific Morning Glory with large blue-violet flowers and three-lobed leaves, *Ipomoea congesta.* The other is a less attractive plant in the Vervain family, as is obvious from the few small purple flowers scattered along the stiff spike, which may arise 5 feet above the ground. It is *Stachytarphata jamaicensis,* or Blue Porterweed.

MAHOGANY HAMMOCK

Also located along the main Everglades National Park Road, this hammock more than lives up to its name by having the largest West Indies Mahogany trees in the continental United States. The champion of all may be seen along the boardwalk, its trunk and limbs covered with air plants and Resurrection Fern *(Polypodium polypodioides).* Many of its companions were victims of Hurricane Donna, the devastating storm that swept across Florida in 1960, and the massive logs that are all that remain of them are being claimed by various kinds of epiphytes. Especially striking are the Long Strap Ferns *(Campyloneurum phyllitidis),* with undivided leaves arching from the base. Each of these fronds resembles an elongated feather, both in its overall shape and in having a series of parallel veins diverging obliquely from either side of the midrib.

Near the entrance to Mahogany Hammock is a colony of Everglades or Paurotis Palm *(Acoelorrhaphe wrightii).* This species grows in clusters, has a very slender trunk, and is one of the "fan" palms; that is, its leaves are divided palmately rather than pinnately. It is a rare Everglades endemic but has been used rather extensively in landscaping.

LONG STRAP FERN
Campyloneurum phyllitidis

SNAKE BIGHT

The Snake Bight Trail branches off of the main Park Road near West Lake and follows an old road south to its namesake (the word "bight" is not misspelled; it means an open bay). Private vehicles are not permitted on this road, and mosquitoes can be annoying even during the winter, so a ride on one of the trams operating out of Flamingo (with insect netting in place) might not be a bad idea. One certain advantage is the highly informative running commentary furnished by a well-versed driver-guide.

Even before the trailhead is reached, the vegetation along the first 7 miles of the Park Road will be found interesting. Papayas *(Carica papaya)* attain heights of up to 20 feet and look like trees with a crown of deeply lobed leaves, but actually they are one of the largest herbaceous plants in the world. Wild, uncultivated individuals like the ones seen here produce fruits, but they are only a few inches long and barely edible.

Another plant of economic importance is *Agave sisalana,* one of the so-called Century Plants and the source of the sisal fiber used in making rope. It was introduced into southern Florida from its native Yucatan.

Conspicuous when in flower is Coral Bean *(Erythrina herbacea),* a large plant in the Legume family. It bears a spike of vivid red, almost tubular flowers up to 2 inches long.

Also along the way is Mrazek Pond, an exceptional place for viewing birds in the winter. The edge of the water is just a few yards from the road, and roseate spoonbills as well as egrets, herons, grebes, and ducks are often seen feeding together close by, while off to one side groups of stately wood storks take up their positions on the shore. The best part is that none of them seems perturbed in the least by the dozen or more photographers and birders with their tripods, telephoto lenses, and spotting scopes.

The trail itself runs along a ridge formed of earth thrown up in the process of dredging a spur canal. The canal is on the left as you enter, and it is on this side that the richest plant life is to be seen. A number of hammock species are there, among them Gumbo Limbo, Mahogany, Strangler Fig, Jamaica Dogwood, and Marlberry *(Ardisia escallonioides),* along with Leather Fern *(Acrostichum danaeaefolium).* Epiphytes include several species of *Tillandsia* and the beautiful Florida Butterfly Orchid *(Encyclia tampensis),* seen on the next page, a summer-blooming species that features a bright magenta blaze on the lip.

Approaching the coastal prairie, mangroves attest to the fact that tidal intrusions have increased the salinity of the water and thereby have made it physiologically less available to plants. The result, paradoxically, is to create a desertlike environment amid wet surroundings. This fosters the growth of cacti, such as Prickly Pear (*Opuntia* spp.) and two species of *Cereus:* one a vigorous climber, strongly three-ribbed, called Barbed-wire Cactus, and the other erect and cylindrical, known as Prickly Apple Cactus. It also encourages salt-tolerant

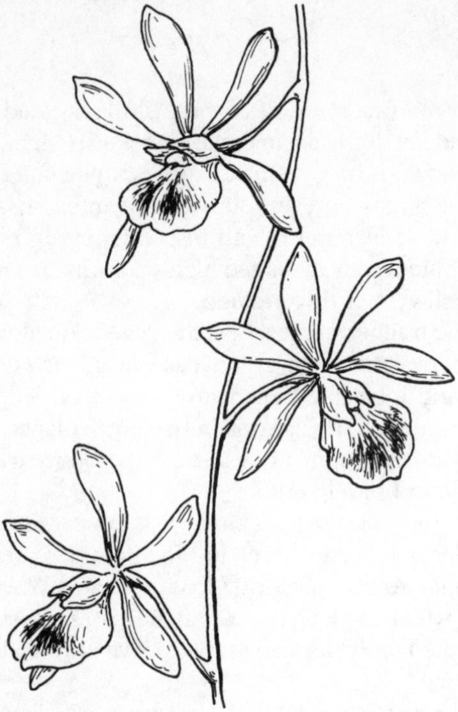

FLORIDA BUTTERFLY ORCHID
Encyclia tampensis

succulents, which are able to conserve water in their tissues. Especially abundant here are Pickleweed and Glasswort.

MATHESON HAMMOCK

This county park is adjacent to the Fairchild Tropical Garden in Coral Gables. To reach it from U.S. 1, turn east on Granada to Cartagena Plaza, then right on Old Cutler Road for about 2 miles. There are parking areas on both sides.

All of the recreational facilities are east of the road on Biscayne Bay, with only the nature trails opposite. This is a dense hardwood hammock rich in native and Caribbean woody plants. In addition to a network of trails, an old road bisects the tract, emerging shortly from the forest onto an old *allée* of Royal Palms—but not before passing through a grove of Cajeput trees *(Melaleuca quinquenervia).*

Prominent among the native trees are the bright green leaves of Wild Coffee *(Psychotria undata).* The veins are so deeply indented as to give them a beautiful "quilted" look reminiscent of the Gardenia, which also is in the Madder family. Although *Psychotria* is related to the true coffee tree and has somewhat similar berries, it cannot be used as a substitute for that beverage, despite the common name.

Another interesting tree, seldom found north of here, is Wingleaf Soapberry *(Sapindus saponaria)*. Its pinnately compound leaves with a winged axis and untoothed leaflets make it easy to identify. The orange-brown fruits are filled with a juicy pulp that lathers abundantly in water to make an excellent substitute for soap.

One of south Florida's largest trees is the Mastic *(Mastichodendron foetidissimum)*, and Matheson Hammock is reputed to have the country's biggest specimen. These trees have straight trunks and are valued for their timber. They bear small yellow fruits that are said to be pleasant tasting but have the somewhat questionable property of causing your lips to stick together.

A large number of interesting ferns thrive in this moist, shady hammock. In addition to the usual two species of *Nephrolepis*, there are Fan Maidenhair *(Adiantum tenerum)*, a species considerably depleted in the wild by collecting, and the alien Chinese Brake *(Pteris vittata)*, to cite only a couple.

Tourists visiting the Fairchild Tropical Garden and learning that picnicking is not permitted there will find it useful to know that the facilities of Matheson Hammock County Park are available next door.

JACK ISLAND STATE PRESERVE

Considering its location northeast of Lake Okeechobee, it comes as a surprise that the flora of Jack Island State Preserve includes a number of species usually associated with mangrove swamps and subtropical hammocks farther south.

The tract fronts on the Indian River, and a sign on the west side of S.R. A1A just above Fort Pierce points to the entrance. There are several miles of trails, and a good choice would be the one that crosses the preserve to an observation tower.

Almost immediately the eye is greeted by the glistening white fruits of Snowberry *(Chiococca alba)* and the pretty white flower heads and amethyst-colored fruits of Florida Sage *(Lantana involucrata)*, see p. 142, against a backdrop of mixed foliage that includes a good amount of Groundsel Tree *(Baccharis halimifolia)*. Tiny yellow blossoms of Creeping Cucumber *(Melothria pendula)* dangling from their climbing vines furnish a nice accent.

Gumbo Limbo trees are scattered throughout, not especially large but unmistakable. There are a number of small Strangler Figs that obviously have sprung from seeds that germinated in the ground. They are easily identified by their large glossy leaves and sessile fruits, some of which are obligingly situated below eye level. In winter, a leafless shrub with branches bristling with greenish

FLORIDA SAGE
Lantana involucrata

yellow apetalous flowers turns out to be Wild Olive, or Florida Privet *(Forestiera segregata).*

The trail is wide and sunlit, and this encourages a variety of small plants to grow in the drier areas. These include magenta-pink Tasselflowers (*Emilia* sp.), Wild Poinsettia *(P. heterophylla),* and a vine known as Crab's-eye *(Abrus precatorius).* The most conspicuous feature of the latter is the fruit: bright red seeds with a single black spot, which are revealed when the pods dry and split open. These have been used in rosaries (another name is Precatory Bead), and at one time were made into children's jewelry—unwisely, since they are so toxic that a single seed can kill a grown man if ingested.

A striking little plant is *Rivina humilis,* Rouge Plant or Baby Pepper, with spikes of pink flowers and little scarlet berries often appearing simultaneously. Something about its general aspect betrays its relationship to the Pokeweeds (*Phytolacca* spp.) despite many differences in the characters of the respective genera.

Red and Black Mangroves as well as Buttonwood are just about everywhere, sometimes behind a thin screen of trees but often right along the roadside. Nearby on firmer but moist ground are patches of several succulents: Perennial Glasswort *(Salicornia virginica),* Saltwort *(Batis maritima),* and Pink Purslane *(Portulaca pilosa).*

As might be expected, the preserve has interesting bird life, and wood storks can sometimes be seen from the observation tower.

COLLIER-SEMINOLE STATE PARK

Although this facility includes a 4,760-acre mangrove wilderness (where limited exploration by canoe is permitted), cypress swamps, salt marshes, and pine flatwoods, its most popular natural feature is a self-guiding nature trail through a pleasant, shady tropical hammock. The hammock was originally named "Royal Palm," and even today it is one of only a few sites where the native *Roystonea elata* continues to flourish.

One of the first herbaceous plants to be seen, and the most abundant, is Crimson Dicliptera *(Dicliptera assurgens),* a branched perennial with curved, two-lipped flowers. Another is *Alysicarpus vaginalis,* known as False Moneywort, which has lavender papilionaceous flowers and leaves with only one segment.

Many of the predictable woody plants are there, including Wild Coffee, Jamaica Dogwood, Gumbo Limbo, and Pond Apple. White Stopper *(Eugenia axillaris)* can be detected by its faint skunky odor, but only within a distance of a few feet.

In addition to Leather and Strap Ferns, there are Golden Polypody *(Phlebodium aureum)* and Shoestring Fern *(Vittaria lineata),* both epiphytes. The last hangs in tufts from the old leaf bases of Cabbage Palms, its long pendent fronds looking like coarse, dark green grass.

A short spur boardwalk to an observation platform leads to a salt marsh environment. Here there are spreading masses of *Lycium carolinianum,* called Matrimony Vine or Christmas Berry, together with Sea Ox-eye *(Borrichia frutescens)* and Glasswort *(Salicornia virginica).*

Returning to the main trail, two unusual plants claim our attention. The first is Arthritis Vine *(Hippocratea volubilis),* a stout woody vine bearing knots along its stem. The other is Devil's-claw *(Pisonia aculeata);* this is a vinelike tree that climbs by means of branched, viciously curved thorns that make the nearby Strangler Figs look much less menacing by comparison. More benign in appearance than either are White Mangrove *(Laguncularia racemosa)* and Buttonwood *(Conocarpus erectus).*

Collier-Seminole is located on U.S. 41 at its junction with County Road 953 from Marco.

BRIGGS NATURE CENTER

This is another place where a single nature trail traverses several different plant communities, and it is placed in this chapter more or less arbitrarily. Briggs Nature Center is located in the Rookery Bay National Estuarine Sanctuary between Naples and Marco Island, and is reached by Shell Island Road from S.R. 951.

A 2,500-foot boardwalk loop begins in a scrub oak community where Myrtle Oak *(Quercus myrtifolia),* Chapman's Live Oak *(Q. chapmanii),* and Scrub Live Oak (*Q. geminata*) are all represented. Other plants typical of this habitat are Rusty Lyonia *(Lyonia ferruginea),* Rosemary *(Ceratiola ericoides),* and Reindeer Moss (*Cladonia* sp.).

Shortly the trail enters a Slash Pine–Saw Palmetto complex with hammock vegetation. Conspicuous among the latter is Red Bay *(Persea borbonia),* here heavily infested by the gall insects whose black disfiguring growths on the leaves so often facilitate instant identification of this species. Other trees include Dahoon Holly *(Ilex cassine),* and Leather Fern is joined by Bracken *(Pteridium aquilinum).*

Abruptly we reach a wet sawgrass prairie flanked by scattered Pond Apple trees, and as we move to brackish water from Rookery Bay the Sawgrass gives way to salt-tolerant Black Needlerush *(Juncus roemerianus),* Red and White Mangroves, and Buttonwood.

The final leg of the trail passes by a shrubby section where we note Gallberry Holly *(Ilex glabra)* and Hog Plum *(Ximenia americana).*

By now it will have become obvious that while hammocks are confined to the warmest parts of Florida, they are quite widely distributed within that region. This will be further confirmed as their appearance on islands such as Sanibel and the Florida Keys is noted in a subsequent chapter.

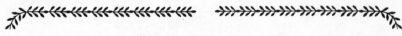

16

Bald Cypress Swamps

ALTHOUGH BALD CYPRESS trees are found in every one of the southeastern coastal states, the Big Cypress Swamp of southern Florida is preeminent, not only because of its size but also owing to its associations with subtropical plants.

Situated northwest of the Everglades, Big Cypress Swamp is defined rather vaguely on most maps. Big it certainly is—2,400 square miles and covering most of Collier County—but it is much more than just a cypress swamp, although those are the predominant trees. Just as in the sawgrass prairies of the Everglades, the precious water is conserved by its own slow rate of flow and the relative imperviousness of the underlying limestone, and nurtures plant communities ranging all the way from freshwater swamps and marl prairies to hammock forests and pine woods.

The durability of cypress wood was too much for the exploiters to resist and, vast though it is, much of the Big Cypress forest has been decimated by lumbering and by fire. There are many trees to be seen, of course, but they are second growth, most of the mature giants having been cut.

Bald Cypress *(Taxodium distichum)* is a tree of ancient lineage, and individuals can attain great age, so it is not surprising to learn that the Giant Sequoias of California belong to the same family. Although it bears cones and has needlelike leaves (see p. 146), the latter are deciduous, hence the name "bald." It often develops a wide buttress-based trunk for stability, as do many other species growing in a water-logged substratum, but is unique in sending up those woody projections known as "knees" above the surrounding water or soil.

There is so much else to see in these swamps, however, that one can almost forget that the Bald Cypresses are there. Take epiphytes, for example—those plants that use trunks or branches merely for support and obtain most of their sustenance from particles of organic matter brought to them by the wind, rain, and dew. They run the gamut from primitive algae and lichens to the most

BALD CYPRESS
Taxodium distichum

highly developed of all plants, the orchids. Of the orchids, the Big Cypress Swamp boasts of no fewer than four native species in the genus *Vanilla,* as well as the alien *V. planifolia* from which commercial vanilla is obtained. This plant had long been used by the Aztecs before it was discovered by Cortés's forces in the 16th century. In more recent times it has escaped from cultivation here, and is fairly common in some places.

But this is only a hint of things to be seen. They say that the way to really experience a swamp is to find a good one and start slogging, so let us consider this method before going on to look at the alternatives.

FAKAHATCHEE STRAND STATE PRESERVE

The immense Big Cypress National Preserve, which is administered by the National Park Service, is open to hiking, but the only marked route within its boundaries is the Florida Trail and conditions are primitive.

Fortunately, the State of Florida has provided facilities for access into the Fakahatchee Strand, which lies just to the west; headquarters are located on the W.J. Janes Memorial Scenic Drive, which leaves S.R. 29 at Copeland. At this point arrangements can be made for a wade trip into the swamp under the leadership of an expert guide. These take place only on certain days of the week, depending upon the season, and advance reservations are essential.

Preparation for a wade is largely mental. First, realize that despite the sinister appearance of swamp water it is *clean*. Then try to put aside those memories of parental admonitions against stepping into puddles. Boots and waders are out; what would you do if they should happen to fill up while you are standing in hip-deep water? Just put on an old pair of sneakers (if they are not old, they will be before the day is over) and long pants, not shorts. Your camera, wallet, and similar articles should be in a waterproof plastic bag.

Once you get your feet wet, there are only a few rules to bother about, and they are based on simple common sense: Before transferring your weight from one foot to another, feel around to be sure you are not stepping onto a slippery submerged limb or hooking your toes under a root (the water is too dark to see through). And skirt any sizable area that is devoid of trees; the water may be too deep for your liking.

Fakahatchee Strand is about 20 miles long and several miles wide (the term "strand" is used locally to denote an elongated forest growing along a slough, or draining channel); and while it still represents an exceedingly rich natural area, there is plenty of evidence of the ravages of past logging, burning, and draining practices. For example, it is crosshatched by a grid of raised roadbeds, or "tramways," which were used to remove timber and on some of which our jeep will now be driven. The square water-filled depressions within this network are the places where we will be spending most of our time on foot. One gets the impression of moving around on a gigantic waffle.

Our first departure from the road is on a tramway too narrow and overgrown for driving. Almost immediately we see gray boles of Florida Royal Palms, scattered survivors of this native species. Ferns are everywhere, and we soon recognize Wild Boston, Swamp, and Strap Ferns. Our destination is the edge of the swamp under Bald Cypresses, where several Water Spider Orchids *(Habenaria repens)* are in bloom. Some are rooted on a piece of floating log, and since the water is virtually motionless all it takes is a touch of a finger to maneuver them into any desired position for photographing. On a dry bank nearby we see Straight Habenaria *(H. odontopetala),* but it is past flowering.

Back on the road we pass ditches filled with Pickerel Weed and the nodding white spikes of Lizard's Tail (*Saururus cernuus*, see the next page), on the way to another site, where we start our first wade.

Bromeliads are very plentiful here, including not only the handsome Cardinal Wild Pine but other members of the genus ranging from the little, relatively rare Hoary Air Plant *(Tillandsia pruinosa)* to the largest species to be found in the United States, the well-named Giant Wild Pine *(T. utriculata).*

Also placed well out of reach of the water, growing on logs and stumps, are Whisk Ferns *(Psilotum nudum).* This species is our only representative of those ancient plants, which are much more primitive than the true ferns and for the most part are extinct. They form upright clumps of rigid, repeatedly branched stems with yellow sporangia.

LIZARD'S TAIL
Saururus cernuus

The little white flowers of Grass-leaved Arrowhead *(Sagittaria graminea)* glitter against the backdrop of dark water and fluted cypress bases, and with our every sloshing step clusters of floating yellow Bladderworts *(Utricularia foliosa)* sway back and forth all around us.

Driving down the tramway once again, we suddenly spot what we take to be a fawn, standing in the road a hundred feet away. As we stare each other down, our guide explains that it is a full-grown Virginia White-tailed deer that belongs to a race midway in size between the typical northern species and the diminutive Key deer. These animals are important as prey to the Florida panther, and we are in the midst of that endangered species' territory. We will, in fact, see Panther Crossing signs along the highway; this does not, however, signify that they are numerous but rather that the survival of the few that do remain is gravely threatened by vehicular traffic.

Our next wade is in a Pop Ash–Pond Apple slough—a plant community frequently encountered in cypress swamps and typified, as the name indicates, by two water-tolerant understory trees, *Fraxinus caroliniana* and *Annona glabra.* At the outset, this presents a startling appearance. Large expanses of the water are covered by a solid-looking film of Duckweed *(Spirodela polyrhiza),* made up of myriads of flat leaflike thalli so small that they can be measured only in millimeters, joined here and there by a patch of Water Lettuce *(Pistia stratiotes).* Out of this pale carpet are thrust masses of bright emerald green lanceolate leaves belonging to Fire Flags *(Thalia geniculata)*—big enough, to be sure, in contrast to the tiny Duckweeds but not even hinting at their ultimate length, which later in the year may reach as much as 3 feet, making them the largest leaves of any native plant except palms. This unusual plant, also known locally as Arrowroot, is rather common in warm, wet areas, and is a relative of cannas and bananas. Less imposing than the foliage are its irregular violet and white flowers, disproportionately small and loosely arranged at the top of a stalk that may reach a height of 10 feet.

Abruptly there is a change in the water surface, which is now populated by *Ricciocarpus natans,* a pretty fan-shaped floating liverwort with the common name of Purple-fringed Riccia.

Fakahatchee Strand is believed to contain more species of orchids than any other place in North America, and this particular slough will prove to be the best spot for epiphytic species that we have seen today, although the ones that are blooming now are among the least attractive. They include the so-called Brown Orchid *(Epidendrum anceps),* which has a very densely flowered raceme of rather pretty but small green blossoms suffused with red, and *E. rigidum* with fewer, quite homely flowers. But then there is the closely related Clamshell Orchid *(Encyclia cochleata),* which takes its name from the showy purple concave lip that is uppermost and has slender yellowish sepals and petals below. *Encyclia tampensis,* known as the Florida Butterfly Orchid, has everything going for it and as a result is the native species most widely cultivated; it is abundant, fragrant, has a long flowering season, and is beautiful in the delicacy of both form and color.

One orchid that has survived in Big Cypress Swamp in spite of heavy collecting is *Cyrtopodium punctatum.* It is called Cow Horn or Cigar Orchid because of the shape of its pseudobulbs, and Bee Swarm Orchid from the appearance of its sprays, which may contain dozens of blooms. Also found here is the Ghost Orchid *(Polyrrhiza lindenii),* surely one of the most striking of all. Its flowers are whitish, and the lip features a pair of very long, twisted lobes that give it an animated aspect and suggest, among other things, a jumping frog (see p. 150). The plant is leafless, and its roots, which look like flattened grayish green worms, fan out over the surface of the support where they anchor the plant and perform photosynthesis as well.

It is always exciting to be present when a new population of an uncommon plant is discovered, and on this occasion it happened to be a species of *Peperomia*—a genus familiar to house plant enthusiasts. There were a great many of these weak-stemmed plants twining around tree trunks, and they seemed different from others that we had seen in that there were multiple flower spikes, although our knowledge was too superficial to be of any further

GHOST ORCHID
Polyrrhiza lindenii

help. Our guide recognized them immediately, though, as *Peperomia glabella*—also known, as you might expect, by the name of Cypress Peperomia.

Big Cypress Bend

It is not necessary, of course, to wade in order to see something of Fakahatchee Strand's swamplands. As a matter of fact, the state has provided the customary accommodation, a raised boardwalk, at Big Cypress Bend on U.S. 41, 7 miles west of the junction with S.R. 29. This structure extends 2,300 feet into the virgin cypress area of Fakahatchee Strand.

Again the vegetation is that of a mixed swamp forest, and includes such diverse plants as Red Maple, Cabbage Palm, and Strangler Fig. The understory is a complex of trees and shrubs such as Flowering Dogwood, Dahoon Holly, Pigeon Plum, Coastal Plain Willow, Myrsine, Marlberry, and Satinleaf, as well as the ubiquitous Pond Apple and Pop Ash. Many of the orchids, bromeliads, and ferns of the central slough are also present here. The boardwalk terminates in a small gator pond lined by Fire Flag, where animal life often presents opportunities for short-range observation and photography.

CORKSCREW SWAMP SANCTUARY

For those who do not care to participate in a wade trip or cannot do so (which would include small children), nothing could be better than a visit to Corkscrew Swamp Sanctuary. Located in the northwestern corner of Big Cypress Swamp, it is reached by driving 16 miles west on County Road 846 from Immokalee. If you approach from the west, it is about 19 miles from U.S. 41; here the route number may be hard to find, but it is also designated Immokalee Road.

Corkscrew Swamp Sanctuary is an 11,000-acre tract maintained by the National Audubon Society, and contains the nation's largest remaining population of virgin Bald Cypress trees, some of them more than 700 years old. Not surprisingly for an organization with an ornithological background, a major factor in its taking measures to prevent the destruction of the tract was the desire to protect an important wood stork nesting area, and these birds are now a leading attraction in the winter.

A self-guiding trail with interpretive signs begins at the visitor center, where feeders attract painted buntings, and proceeds through a short stretch of Slash Pine–Saw Palmetto woods where the trees have never been cut and where red-cockaded woodpeckers can be observed. Two attractive herbs are common here: the little yellow Thimble *(Polygala rugelli)* and False Pennyroyal *(Piloblephis rigida).*

As you approach the start of the boardwalk, there is an open grassy area devoid of trees and dominated at first by Broomsedge (*Andropogon* sp.), with white accents supplied by Long-leaf Violets *(Viola lanceolata)* and Pipewort (*Eriocaulon* sp.).

As the water level increases, Pond Cypresses enter the picture and before long we find ourselves walking through a typical cypress strand. Epiphytic plants are all around, from *Usnea* lichens to *Tillandsias* and *Catopsis berteroniana,* or Yellow Catopsis. Leather Fern *(Acrostichum danaeaefolium)* is plentiful, and there also are specimens of the so-called Golden Polypody *(Phlebodium aureum)* growing on cypress trunks. This unusual fern derives its common name from the long bronze-colored hairs that cover the rootstock and the round sori neatly arranged on either side of each pinna's midvein.

As the swamp becomes still wetter, the cypress trees are seen to be much larger and their foliage feathery instead of appressed as on the Pond Cypresses. These are the typical species *(Taxodium distichum)*—the true Bald Cypresses. Here, too, Leather Fern is replaced by Swamp Fern *(Blechnum serrulatum),* and this is joined by Wild Boston Fern *(Nephrolepis exaltata),* Strap Fern *(Campyloneuron phyllitidis),* and Royal Fern *(Osmunda regalis).* Pond or Custard Apples, Red Bay *(Persea borbonia),* and Virginia Willow *(Itea virginica)* help to make up the understory.

Halfway around the boardwalk a short spur leads to an observation platform situated in a marsh with a willow head nearby. Along the way one of the showiest flowering plants, Swamp Lily *(Crinum americanum),* will be seen in numbers together with two species of Arrowhead, *Sagittaria lancifolia* and *S. latifolia.* Climbing plants include Moon Flower *(Ipomoea alba),* with large white morning glory-like blossoms, and the more delicate Milkvine *(Sarcostemma clausum).*

An unusual flower to be seen here is the Climbing Aster *(Aster carolinianus).* Its pink or purplish flowers are not unlike those of many other attractive Aster species. It is unique, however, in having woody arching stems that clamber over other vegetation and are sometimes several yards in length.

This spur is one of several places from which wood storks can be observed during nesting season. As you walk toward the platform, look for them high up in cypresses to the left. Pileated woodpeckers are conspicuous in this area as well.

At another spur farther along the boardwalk there is an abundance of the poisonous Water Hemlock *(Cicuta mexicana).* One of the most beautiful native flowers growing here is the Wild Hibiscus *(Hibiscus coccineus),* with 6-inch scarlet flowers and leaves divided into long, graceful lobes.

"Lettuce lakes" are a major feature of the Corkscrew Sanctuary; here floating plants of Water Lettuce *(Pistia stratiotes)* cover virtually all of the ponds' surfaces. The light green leaves are invested with silvery hairs, but they are also thick, ribbed, and grow in rosettes, and it is apparently these characteristics that

WILD HIBISCUS
Hibiscus coccineus

suggested a similarity to leaf lettuce. The small flowers possess a spathe and spadix, and are typical of the Arum family to which it belongs.

Logs in and around the lettuce lakes are favorite basking sites for Florida red-bellied turtles, and alligators are frequently seen. Anhingas and white ibis feed actively here, and the taller Bald Cypresses are nesting places for wood storks.

The last part of the boardwalk is notable for the abundance of epiphytes—mostly bromeliads and lichens but also some orchids. The transition from swamp to pineland is quite abrupt, and with it the flowering plants suddenly diminish in size to include such species as Bay Lobelia *(Lobelia feayana),* Blue-eyed Grass *(Sisyrinchium atlanticum),* Lawn Orchid *(Zeuxine strateumatica),* and a white form of Spanish Needles, *Bidens alba* var. *radiata.*

LOXAHATCHEE NATIONAL WILDLIFE REFUGE

On most road maps Loxahatchee National Wildlife Refuge appears to be a featureless area with no evident means of entering or traversing it. Well, it is far from devoid of interest, and much of it can be explored by water. For the pedestrian visitor, there is an entrance road leading west from U.S. 441 between S.R. 804 and S.R. 806 (this is approximately opposite Boynton Beach), and this takes you directly to a fine interpretive center.

A major feature here is the Cypress Swamp Boardwalk, a half-mile loop. This is a long way from Big Cypress Swamp in both size and diversity, but it does provide an excellent introduction to the vegetation of south Florida swamps in general. Some plants that occupy widely separated ecological niches farther south can be found here growing in close proximity. Pond Apple *(Annona glabra),* Buttonbush *(Cephalanthus occidentalis),* Arrow Arum *(Peltandra virginica),* Coastal Plain Willow, and Bald Cypress might all be seen at once without turning the head.

Many of the trees present a startling appearance with their coats of the red Blanket Lichen *Herpothallon sanguineum.* This is the same lichen that forms scabby pink patches on large Live Oaks, for example, but when it grows on smaller straight trees and covers the boles completely, as it often does here, another of its common names, Baton Rouge, which means "red stick," is particularly apt.

Also adorning the trees are bromeliads, air plants in the Pineapple family. The largest is *Tillandsia fasciculata,* called Cardinal Wild Pine because of the showy pink bracts from which the small flowers emerge, and from the resemblance of its foliage to that of a pineapple (see the next page). The so-called Ball Moss *(T. recurvata)* and Spanish Moss *(T. usneoides)* are the only two species in which plants do not grow individually but are intertwined; in the former they are short and curled, forming a roundish cluster, while the long stems of the latter hang in gray festoons.

CARDINAL WILD PINE
Tillandsia fasciculata

There is a generous assortment of ferns here also. These include Royal Fern *(Osmunda regalis),* which occurs far to the north as well, but the others are limited to Florida and the tropics: Leather Fern *(Acrostichum danaeaefolium),* our tallest fern, sometimes exceeding 10 feet in height; Sword Fern *(Nephrolepis biserrata);* the similar Wild Boston Fern *(N. exaltata),* from which many horticultural forms were developed; and Swamp Fern *(Blechnum serrulatum).* There is even a tiny floating fern, *Salvinia rotundifolia,* called Water Spangles, with nearly circular hairy fronds less than a half-inch long.

Since cypress swamps are not the best places in which to find showy flowers, it is difficult to resent the intrusion into the area of the South American Primrose Willow *(Ludwigia peruviana),* a large plant that goes a long way toward brightening these dark places with stunning four-petaled yellow flowers up to 2 inches across.

The Marsh Nature Trail is also in this part of the refuge—actually a 0.8-mile walk around an impoundment (visitors are welcome to walk on *any* of these dikes, for that matter). It is primarily intended for wildlife observation, and Florida soft-shell turtles, fulvous whistling ducks, and yellow-billed cuckoos are among the great many species that are to be seen.

Common aquatic plants are Cattails (*Typha* sp.), Arrowhead *(Sagittaria falcata),* Spatterdock *(Nuphar luteum),* and Water Lily *(Nymphaea odorata).* Water Lettuce *(Pistia stratiotes)* forms large mats for common moorhens to stroll across, and its gray-green leaves form a camouflaging blanket on the backs of the alligators when they surface in the middle of a patch of these plants.

In slightly drier thickets there is Florida Elder *(Sambucus simpsonii).* Planted at intervals around the dike are several clusters of assorted trees, including Coconut Palms *(Cocos nucifera),* Coco Plum *(Chrysobalanus icaco),* and West Indies Mahogany *(Swietenia mahagoni).*

There are no picnicking facilities in the refuge, but just 6 miles south on the right-hand side of U.S. 441 is Pinewoods County Park, which has picnic tables in the shade of Slash Pines. For those who find it difficult to eat outdoors without botanizing, there are such treats as Beauty-berry *(Callicarpa americana),* Caesars Weed *(Urena lobata),* and species of *Emilia* and *Sida* to savor.

17

The Florida Keys and Other Islands

SO MUCH OF THE south Florida coast has been developed, or is otherwise inaccessible, that there would hardly seem to be enough examples of natural seaside areas available to warrant devoting a chapter to them. Indeed, this might well be the case if it were not for the Florida Keys.

Unlike the beaches farther north, which are composed principally of glittering particles of quartz, the snow white strands of southern Florida are of organic origin. They remind us that this land was covered intermittently by the sea until fairly recent times, and that these waters were populated by marine animals such as we now find on and around coral reefs. As these creatures died, their limey shells and skeletons accumulated on the seabed to form what became the floor of Florida when it finally emerged for the last time. At Cape Sable and in the Keys, the proportion of quartz is negligible and the sands are almost totally derived from corals, mollusks, foraminifera, and the like.

The succulents and other small plants of the dunes are pretty much like those found on the more temperate Florida beaches, and hold few surprises. It is among the shrubs, vines, and trees that we see numerous species of West Indian origin, reminders that we are in the nearest thing to a tropical environment that the continental United States has to offer.

THE FLORIDA KEYS

The Florida Keys are strung out like a path of stepping stones from just below Miami to the Dry Tortugas. The major part of the chain is easily accessible, thanks to the Overseas Highway—the final lap of U.S. 1—which takes you from Key Largo to Key West in about 3 hours. Points of interest are easily located by means of milepost markers, mile 0 being at Key West.

The first link to join the Keys from one end to the other was an extension of the Florida East Coast Railroad, built by Henry Flagler, an associate of John D. Rockefeller, between 1904 and 1912. Following massive damage by a severe hurricane in 1935, it was abandoned, and plans were made for a highway to be built over the old roadbed. The present highway is a marvel of construction, and its contribution to the development of the Keys has been incalculable. It is well to remember, though, that essentially this is the equivalent of a hundred-mile-long two-lane backcountry road in that it affords few opportunities for avoiding bottlenecks resulting from accidents or road repair work. Even a minor collision can keep you sitting in your car for several hours. And if you have ever wondered what the middle of nowhere looks like, you have only to look out of the window.

In many places where the highway appears squeezed between the Atlantic Ocean and Florida Bay there is an obvious dearth of trees. Utility poles furnish an alternative for the osprey, however, and while driving along you frequently can spot these stately birds on their immense nests, which are built principally of sticks. Double-crested cormorants perch on the wires in considerable numbers but seem to prefer the sloping cables that guy up the poles, where it must be much more difficult to maintain a purchase with their ill-adapted feet.

Geologically, the long thin line of the Upper Keys is what remains of an old coral reef that formed offshore when the main land mass ended somewhere around Sebring, but it is now pitted and eroded into jagged rock. The Lower Keys, like the Everglades, are of Miami oolite, a limestone made up of material that drifted to the bottom of the shallow sea that covered all of what is now southern Florida.

Lying farther south than the mainland, the Keys are even more suited to tropical growth (they almost never experience frost). This has led to the importation of ornamental shrubs and trees from many parts of the world, and there have been large plantations of pineapples and coconuts at various times. The little yellow Key Lime *(Citrus aurantifolia)* is of course named for the Florida Keys, and the thorny trees are well adapted to growing in the sparse soil overlying the coral rock. They are not native, however, but trace their origin to seeds brought from the East Indies by Spaniards four centuries ago.

Long Key

The name Long Key has several significant associations with the railroad. During construction of the 12,000-foot-long viaduct a violent hurricane struck, causing extensive damage and killing more than a hundred workers. Later it became the site of Flagler's Long Key Fishing Club, which was to attract the world's most famous sports fishermen.

Today it boasts a fine nature study facility, the Long Key State Recreation Area. The entrance is between mile markers 68 and 67, on the south side of the highway. A nature trail loops through several types of environment, beginning with a small hammock where a placard considerately points out a Poisonwood tree *(Metopium toxiferum).* It explains how the sap, which can produce severe blistering of the skin upon contact, frequently seeps through the outer bark, causing distinctive dark blotches. While this is very useful in recognizing and avoiding contact, becoming familiar with the glossy, yellow-edged foliage is also advisable, and this is especially important since Poisonwood is plentiful all along this trail.

The path soon emerges into an open area with packed sand underfoot and scattered shrubs and small trees. Conspicuous among them are the round, scurfy brown fruits of Wild Sapodilla *(Manilkara bahamensis).* The true Sapodilla, whose latex is used to make chicle, the basis of chewing gum, is a closely related species, *M. zapota.*

Climbing over some of the plants are the tangled yellow stems of Love Vine *(Cassythia filiformis),* which might easily be taken for one of the Dodders (*Cuscuta* spp.). Looking somewhat like a vine but actually a sprawling shrub is *Ernodia littoralis,* called Golden Creeper not because of its tubular flowers (which may vary from white to red) but because its succulent leaves tend to take on a yellowish cast.

The trail eventually leaves these sunny spaces for the shade of Australian Pines *(Casuarina equisetifolia)* growing on a sandy berm a little distance back from the ocean. Here we find the handsome, slightly irregular blooms of *Cassia chapmanii,* and the pretty yellow flowers with bunches of orange stamens belonging to *Mentzelia floridana.* There are at least two popular names for the latter, Stickleaf and Poor-man's-patches, both attributable to the fact that the stems are clothed in hairs that are armed with minute barbs.

An interesting small tree occurring here is the Darling Plum *(Reynosia septentrionalis),* a member of the Buckthorn family. It has dark oblong leaves notched at the end, and produces edible half-inch purple fruits with a spiny tip.

As the canopy opens up again, the number of species increases greatly. There are Lantanas (both *L. camara* and *L. involucrata*), Dayflowers *(Commelina erecta),* Purslane *(Portulaca oleracea),* and Scorpion-tail *(Heliotropium angiospermum).*

Shrubs include the unusual Seven-year Apple *(Casasia clusiifolia);* in the West Indies it is called Genipa, which also was its former generic name. The fruit is lemon shaped and green at first, then yellow with brownish spots, but is not ripe or edible until it shrivels and turns black like a prune. This takes an exceptionally long time, but certainly not anything approaching 7 years.

Still farther along the dune we find an attractive little composite with heads consisting entirely of white tubular flowers and black stamens projecting from them. This combination has given it the name Salt-and-Pepper, and the scientific label *Melanthera hastata.*

Near the end of this sandy trail there are large patches of the narrow-leaved Wild Poinsettia *(P. pinetorum)* and masses of the showy Morning Glory, *Ipomoea congesta.* The path then changes to a boardwalk overlooking a mangrove-bordered tidal lagoon and terminates at an observation tower.

Big Pine Key

In 1954 the federal government established a refuge here to protect the indigenous Key deer, which had suffered through excessive hunting and loss of habitat almost to the point of extinction. These diminutive animals belong to a race of the Virginia white-tailed deer, adult bucks measuring no more than 28 inches at shoulder height.

To reach the refuge, turn north from U.S. 1 on Key Deer Boulevard, which is between mile markers 31 and 30. At 3 miles, on the left-hand side, is an old quarry known as Blue Hole. Here, as well as in natural depressions on Big Pine Key, the fresh rainwater that is so essential to the deer's survival collects. One-quarter mile farther is the Jack C. Watson Nature Trail, a two-thirds-mile interpretive loop with emphasis on the pinelands and hammock vegetation; an excellent descriptive booklet is available at the trailhead.

Initially, the principal trees are South Florida Slash Pine (*Pinus elliottii* var. *densa*) and Cabbage Palm *(Sabal palmetto),* but farther along these are supplanted by Brittle Thatch Palm *(Thrinax microcarpa)* and Silver Palm *(Coccothrinax argentata)* along with the low-growing Saw Palmetto *(Serenoa repens).*

Understory shrubs and trees include Poisonwood, Snowberry *(Chiococca alba),* Wax Myrtle *(Myrica cerifera),* and Sweet Acacia *(A. farnesiana).* Fairly common is Locust Berry *(Byrsonima lucida),* which has clusters of small flowers with five fan-shaped petals each abruptly narrowed to a claw. These are white at first but quickly turn to yellow or deep rose, and a single bush may display several different colors at one time.

In addition to Bracken (*Pteridium aquilinum* var. *caudatum*), two interesting ferns occur here. One is Plumy Brake Fern *(Pteris longifolia),* which is found only in south Florida and tropical America. The other has the odd name of *Anemia adiantifolia* and is known as Pineland Fern. The genus has been placed in the Climbing Fern family, but this species is erect and quite short. Its structure is unusual in that it consists of a pair of long fertile fronds overtopping a single triangular sterile frond.

Among the few flowering herbs in this comparatively dry area is Pineland Croton *(C. linearis),* a little plant with stems and leaf undersides covered with a brown suedelike pubescence. Vivid color is supplied by a vine, *Jacquemontia pentantha,* which flaunts bright blue three-quarter-inch flowers that look like flat morning glories and are admiringly called Blue Hats.

White-top Sedge *(Dichromena colorata*, shown on p. 160) along the edges of the path tell us we are approaching a wetter environment, and this is confirmed by a grove of Buttonwoods. Buttonwood—or rather the charcoal made from it—was for many years the principal fuel in the Keys.

As we come into the hammock area we see what is certainly one of Florida's most attractive trees. It is a small evergreen called Satinleaf, or *Chrysophyllum oliviforme,* and both its common and generic names derive from the fact that the lower surfaces of its leaves bear long copper-colored hairs that feel like satin when stroked with the finger.

WHITE-TOP SEDGE
Dichromena colorata

Lignumvitae Key

When you consider that no fewer than 42 bridges were built to carry the Overseas Highway from one key to another, it may seem unfair that the state's one official botanical site should be accessible only by boat, but the geographic placement of Lignumvitae Key in Florida Bay off the southern end of Islamorada provided no choice. Tour boats do run from the Indian Key fill site at mile 78.5, but because of the infrequent scheduling, the need for making reservations in advance, and the difficulty of keeping to one's timetable in the Keys, it might be advisable to negotiate the rental of a boat at a nearby marina.

Much of the value of the place lies in the fact that it largely escaped being stripped of its hammock-type vegetation during the days of frenetic development. It is named for Holywood Lignumvitae *(Guaiacum sanctum),* a small evergreen tree with unusual blue flowers. Its wood, which is so dense that it sinks in water, is one of the hardest known and has an extremely high resin content, properties that adapt it to a number of specialized uses. This has brought about its virtual disappearance from the Keys, although it is widely distributed in tropical America.

Opposite Lignumvitae Key on the ocean side is Indian Key, where Dr. Henry Perrine experimented with the growing of a great many exotic plants. He was killed there in 1840 when a band of Calusa Indians led by Chief Chekika raided the island.

Key West

The Overseas Highway—and for most tourists the Florida Keys—ends at Key West, only 90 miles from Havana. The recorded history of Key West goes back to Ponce de Leon's voyage of discovery in 1513. By turns, it has been famous as a pirate hangout and for ship salvage, salt evaporation, turtle meat, sponge fishing, cigars, and shrimp. In recent times it has served as a naval base and as a second White House, and now the emphasis is on tourism.

The location of Key West makes it possible for you to stay at the southernmost motel in the country, photograph the southernmost house, have a drink at the southernmost bar, and so on. But it also has permitted residents to surround their houses with exotic plants: trees like Royal Poinciana *(Delonix regia),* originally from Madagascar; shrubs such as the lovely but very poisonous Oleander *(Nerium oleander)* from the Mediterranean area; and vines including Paper Flower (*Bougainvillea* sp.), a native of Brazil.

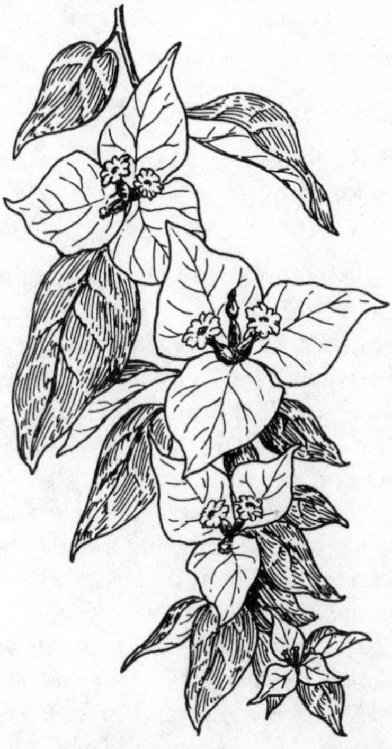

PAPER FLOWER
Bougainvillea sp.

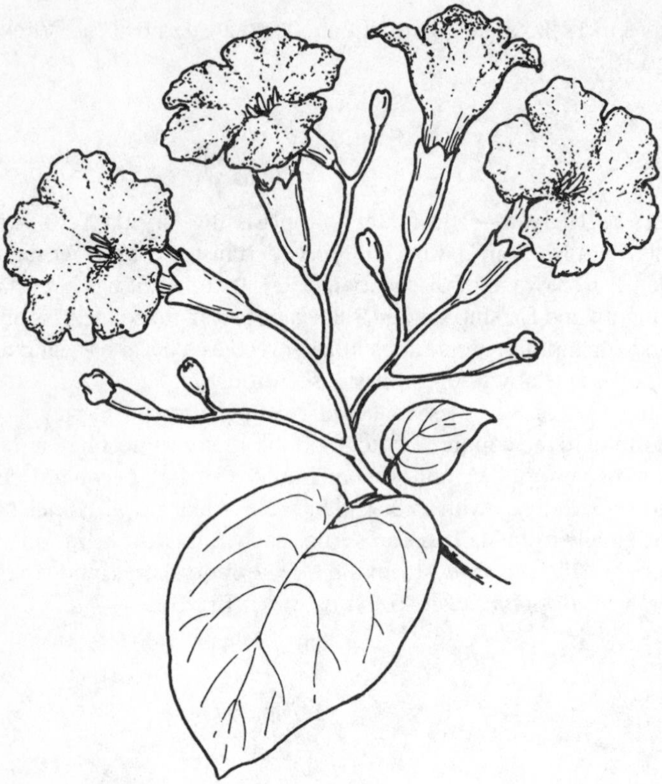

GEIGER TREE
Cordia sebestena

Two residences in Key West that are open to the public are in the midst of lush tropical verdure, but perhaps are more significant for their historical associations than for their botanical interest. They are the Ernest Hemingway House and Gardens, at 907 Whitehead Street, where the author lived for many years, and the so-called Audubon House—actually the home of Captain John Geiger, a harbor pilot and "wrecker," and now converted to a museum commemorating John James Audubon—at 205 Whitehead Street. There is a common belief about the Geiger Tree *(Cordia sebestena)* that holds that Audubon was responsible for naming it for his friend, whom he visited during his brief but productive stay in Key West, and that this accounts for the existence of a young specimen at the entrance to the Geiger house. It now seems, however, that the name was used by the English naturalist Mark Catesby in a work published before Audubon was born. None of this need detract from one's visual enjoyment of *Cordia sebestena,* of course. It is a handsome tree with large oval evergreen leaves and clusters of bright orange, trumpet-shaped flowers more than an inch across.

SANIBEL ISLAND

Two well-known Americans are responsible for having firmly inked in little Sanibel Island on the natural history maps: Anne Morrow Lindbergh, writing symbolically of its seashells in her poignant book *Gift From the Sea,* and J. N. Darling (or "Ding," as he signed his famous political cartoons), whose zeal for conservation led to the establishment of the national wildlife refuge bearing his name.

Sanibel is a subtropical island opposite Fort Myers, and is connected to the mainland by a 3.5-mile causeway and toll bridge from Punta Rassa via S.R. 867. A short bridge at the western end joins it with its smaller neighbor, Captiva Island.

The beaches are justly famous the world over for the phenomenal abundance and the variety (more than 400 species) of their seashells. Most aficionados are perfectly content with those virtually covering the 14-mile beach, worn and faded though they may be, while secretly hoping for a storm of several days' duration to pile up on the beach mounds of new and perhaps rare specimens from the vast offshore beds. Perfect shells freshly taken from the living animals can be obtained by digging in the mud flats (a practice that is not condoned) or they may be purchased from dealers, in which case they may have been dredged from far distant seas.

Because of its natural appeal, Sanibel today finds itself a crowded and highly developed resort, but on the whole it is tastefully planned. There is limited public access to the beaches, where two of the most conspicuous plants are Sea Grape *(Coccoloba uvifera)* and, especially when it is in bloom, Spanish Bayonet *(Yucca aloifolia).* Railroad Vine *(Ipomoea pes caprae),* a beautiful purple-flowered morning glory, sends its runners across the sands in perfectly straight lines for many yards, and it is easy to see the resemblance between the parallel rows of these vines and railway tracks. The specific epithet is equally obvious, an allusion to the rounded leaves with indentations at both ends, suggesting the cloven hooves of a goat.

The area around Sanibel Lighthouse at the eastern tip of the island is a favorite spot for picnicking, beachcombing, and botanizing. Also popular are the parking areas along the causeway, where nesting black skimmers and other birds can be observed. Here can be found sand-loving plants like *Crotalaria pumila,* one of the Rattleboxes, and the pretty Seaside Gentian *(Eustoma exaltatum).*

Landscaping in the town is lush and colorful thanks to the use of exotics like Golden Trumpet *(Allamanda cathartica);* white, pink, and red Chinese Hibiscus *(Hibiscus rosa-sinensis);* Mahoe *(H. tiliaceus)* with yellow flowers that quickly turn deep orange; Poinsettia *(Poinsettia pulcherrima);* and species of *Ixora* and *Bougainvillea.* Mistakes have been made, of course, and there are many more Australian Pines, Cajeput trees, and Brazilian Peppers than anyone bargained for. But there are forces working hard to conserve the island's natural heritage, and the tangible products of their efforts are a delight to every concerned visitor.

J. N. "Ding" Darling National Wildlife Refuge

The "Ding" Darling Refuge is literally for the birds, and they may be seen from your car along the 5-mile one-way Wildlife Drive and from its observation tower, or you can rent a canoe at the marina and explore by water. Incidentally, the dike on which you drive was originally built for mosquito control (they still are a problem here during the summer months) but now assists in manipulating water levels to accommodate migrating waterfowl.

Roseate spoonbills can be seen in large numbers, especially early or late in the day. Common sights are brown pelicans looking precarious as they perch on the tops of mangrove trees, and anhingas spreading their wings in the sun to dry. Alligators are quite numerous, and their feeding around dusk is frequently advertised by loud splashes. The appealing little short-eared marsh rabbits are quite tame and unconcernedly share the footpaths with humans.

Most of the refuge is Red Mangrove habitat, but by far the most interesting area botanically is a sand ridge hammock near the end of the drive, served by the Gasparilla Trail, a one-half mile self-guiding loop. The entrance to the trail is encircled by an interesting assortment of trees typical of the region: Coconut Palms, Strangler Figs, Australian Pines, Sea Grapes, and a Royal Poinciana *(Delonix regia).*

The first few steps bring you to a large stand of Night-blooming Cereus *(Cereus undatus)* clambering over *Sabal Palmettos.* Nearby is a bed of *Sansiveria,* a big relative of the Snake Plants commonly grown in pots up north, with 2-foot spikes of flowers, and a treelike Spanish Bayonet *(Yucca aloifolia).* A number of species generally found in tropical hammocks throughout southern Florida are here: Wild Coffee, Gumbo Limbo, Snowberry, Nicker-bean, and an especially large Mastic Tree (*Mastichodendron foetidissimum).* There are Key Lime trees, left over from an old citrus grove, as well as Lime Prickly Ash *(Zanthoxylum fagara),* also known as Wild Lime. The trail ends in a thicket of Black Mangroves where raccoons are often seen foraging even in the daytime.

The refuge also administers the smaller and very different Bailey Tract, located on the Gulf side of the island, one-half mile south of Periwinkle Way on Tarpon Bay Road. Numerous ponds are interlaced with trails and the attraction—other than the bird life, of course—is the proliferation of herbaceous wildflower species along the grassy sunlit paths.

An extraordinarily pretty combination is made up of showy Marsh Pinks *(Sabatia grandiflora)* mixed with the little sky blue Bay Lobelia *(Lobelia feayana).* Along with these are several composites: Yellowtop *(Flaveria linearis),* Daisy Fleabane *(Erigeron strigosus),* and the omnipresent white Spanish Needles; Lawn Orchids, Gaura *(Gaura angustifolia),* Common Nightshade *(Solanum americanum),* and Narrowleaf Ground Cherry *(Physalis angustifolia).*

Wetter areas contain Capeweed *(Lippia nodiflora),* White-bracted Sedge *(Dichromena latifolia),* and Saltwort *(Batis maritima),* with Southern Cattails

(Typha domingensis), and Water Lilies *(Nymphaea odorata)* in some of the ponds.

Sanibel-Captiva Conservation Foundation Nature Center

Less well-known than the Ding Darling Refuge—but definitely not to be missed—is the nature center established by the Sanibel-Captiva Conservation Foundation for education, research, and preservation. The entrance is on the south side of the Sanibel-Captiva Road.

In addition to an interpretive center and a native plant nursery, there is a network of trails totaling 4 miles in length, and walks may be taken either alone or with a guide. To take one example, the Elisha Camp Trail is reached by a boardwalk over a wetland swale dominated by Cordgrass *(Spartina bakerii).* It then traverses a shell ridge.

One of the most striking native species is Coral Bean, named *Erythrina herbacea* but in southern Florida almost always a woody plant. Even stranger than the blunt three-lobed leaflets are the long tubular flowers of bright crimson.

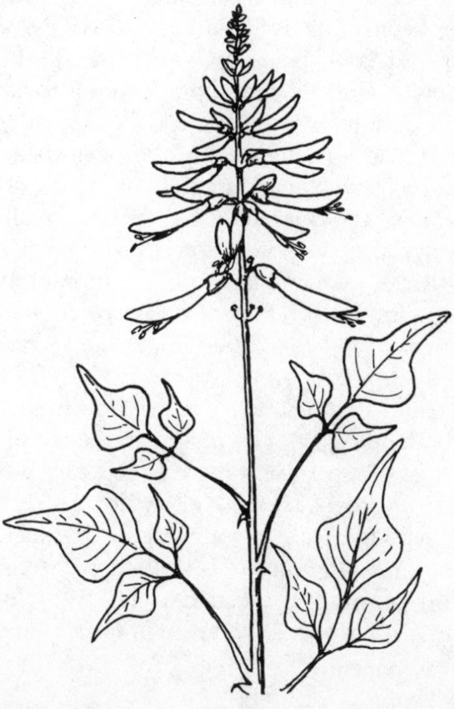

CORAL BEAN
Erythrina herbacea

This area is replete with interesting plants, many of which already have been noted elsewhere, but a few of the others deserve mention. White Indigo Berry *(Randia aculeata)* produces white berries suggestive of Snowberry, but surprisingly the inside is filled with a juicy blue-black pulp surrounding the seeds. Another shrub with odd fruits is Cat Claw *(Pithecellobium unguiscati),* a legume with grotesquely twisted pods; when these split open, black seeds can be seen hanging from threads of red pulp. In Varnish Leaf *(Dodonaea viscosa)* it is the shiny, resin-coated leaves that attract attention.

BLOWING ROCKS PRESERVE

Jupiter Island, which lies off the east coast opposite Jonathan Dickinson State Park, is an important sanctuary for sea turtles, of which at least three species nest there. Near the southern end of the island is Blowing Rocks Preserve, maintained by The Nature Conservancy for the protection of these reptiles but open to the public. Much of its popularity may be due to the fact that its sandy beach is enhanced by outcroppings of Anastasia limestone, eroded by the surf into grotesque "moonscapes."

The entrance path is through a grove of Australian Pines *(Casuarina equisetifolia)* and leads to a trail that goes along the top of a dune and affords a fine opportunity to study some characteristic seaside plants of southern Florida.

In this region of continuous salt-laden breezes very few plants can achieve even moderate height. One is Sea Grape *(Coccoloba uvifera),* which is remarkably salt tolerant. Sea Grapes cannot be picked in bunches, as the individual fruits ripen a few at a time; the most practical way to harvest them is to shake the tree over a drop cloth. They make a delicious jelly, though, and when one realizes that the plant is a member of the Buckwheat family, the image of hot buckwheat pancakes with sea grape jelly is inescapable.

Another is *Suriana maritima,* or Bay Cedar, a small tree with tenacious roots and thick, downy leaves. Its yellow flowers are attractive but sparsely distributed; old yellowing leaves often create the impression of more blossoms than there actually are. A showier shrub is Sea Ox-eye *(Borrichia frutescens),* a fleshy composite with yellow flowers found near the seacoast throughout the southeastern states.

The other plants are prostrate, or at least low in stature, and grow just behind the crest of the dune. Many of them are armed, the most obvious being Prickly Pear Cactus (*Opuntia* sp.). Then there is Puncture Weed *(Tribulus cistoides),* a creeping introduction with bright yellow flowers set off against finely pinnate foliage and seed pods equipped with very sharp spines. The widespread Spurge Nettle, or *Cnidoscolus stimulosus,* is clothed with stinging hairs, and with good reason has also been given the name of Tread-softly. It is, nevertheless, an attractive plant with white salverform flowers formed by the deeply lobed calyxes.

Growing in large patches, the Beach Sunflower *(Helianthus debilis)* is the most conspicuous flowering plant here. The flowers have large reddish brown centers surrounded by broad yellow rays, and keep turning in unison to face the sun. Its sprawling growth habit, unique among the sunflowers, makes it well suited to this environment. Another rather lax plant, with delicate straw-colored bell-shaped flowers, is the Seaside Ground Cherry *(Physalis viscosa).*

One of the more vigorous inhabitants of this sandy habitat is the native Beach Bean *(Canavalia maritima).* This is a robust trailing vine with large trifoliate leaves and rose pink papilionaceous blossoms followed by 4-inch-long pods.

To reach Blowing Rocks, turn east from U.S. 1 onto S.R. 707 at Tenesqua and look for the Nature Conservancy parking lot on the right.

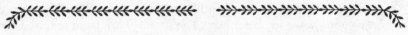

18

Great Lakes Dunes and Forests

THE FIVE Great Lakes, aggregating more than 94,000 square miles in area, not only are the largest in the United States but qualify individually for listing among the 15 biggest in the entire world. They were created when the last of the continental glaciers, which had repeatedly overwhelmed the land during the Pleistocene epoch, finally receded leaving huge water-filled depressions. Also left behind by the wasting ice and its meltwater streams were great accumulations of debris, much of it worn down and crushed by the massive glacier during its long advance. These deposits formed the basis for sandy shores and ultimately, through the agency of prevailing winds, the formation of dunes.

Onshore winds pick up loose sand from the beaches and carry it inland, but whenever it meets an obstruction its migration is slowed and a fore dune begins to build up. Certain grasses with a tolerance for blowing sand are able to gain a foothold here and thus prepare the way for additional vegetation. Should anything disturb this stabilizing plant cover, however, the sand at the edges of the dune will start to cave in and winds will hollow it out to form a bowllike "blowout."

Anyone who has seen the fury of a Great Lakes storm is likely to question how any plant life could possibly exist on the exposed beaches, but there are at least two xerophytes that doggedly insist upon reappearing every year. The radiating stems of Seaside Spurge *(Euphorbia polygonifolia)* help it to hang on by forming prostrate mats flat against the sand, permitting the drying winds to pass harmlessly over it. Sea Rocket *(Cakile edentula),* a more erect herb with tiny lavender flowers, faces the onslaught but has deep anchoring roots and is protected from desiccation by having succulent stems and leaves. Both plants are also found on the tempestuous Atlantic seacoast, but there as well as here they are transitory.

For permanent vegetation it is necessary to move back to the point where clumps of grasses have become established and, by trapping the blowing sand, have initiated the formation of dunes. The most effective of these sandbinders is Marram Grass *(Ammophila breviligulata),* which not only is able to spread by means of vertical rhizomes but can adapt to periods of dryness by rolling up its ribbed leaf blades to reduce transpiration.

Tall Wormwood *(Artemisia caudata)* is a grayish composite with narrowly dissected leaves and numerous small flower heads, frequently growing near the fore dunes. If you find it you might also look for Clustered Broom-rape *(Orobanche fasciculata),* a small plant bearing purplish flowers but devoid of green color. This is a root parasite on various species but in this area seems to favor Tall Wormwood as a host.

Finding Sand Cherry *(Prunus pumila)* along the beach is cause for surprise in a number of ways. For one thing, it obviously is thriving in a habitat that does not look as though it would support any woody plants. A decumbent shrub with narrow leaves, its general appearance suggests some kind of willow. And the juicy, dark purple fruits are about three-eighths of an inch in diameter—seemingly too large for a native cherry.

The progression of dune-stabilizing plants leads inevitably to trees, and one of the first to appear is the Eastern Cottonwood *(Populus deltoides).* In much the same way that Marram Grass sends up new shoots when buried by the sand, this indomitable tree responds by putting out additional roots from its trunk. As the dunes become more stationary, other plants arrive, the soil is built up, the winds are buffered, and the foundation is laid for succession to one or another of several forest communities.

Some of the most interesting forests in the Great Lakes region will be found in glaciated Wisconsin and Michigan. Sandwiched between the coniferous boreal woods, which lie mostly in Canada, and the mixed hardwoods that take up much of the eastern central United States, this belt is known as the Transition Forest and becomes a mixture of both as trees from the adjacent zones extend their ranges southward and northward. Within it, variations in composition occur where certain species achieve a concentration in response to differences in soil or climate.

A few of the trees, especially some that are not commonly encountered farther south, are worthy of mention. Larches are our only northern deciduous conifers, and around the Great Lakes the genus is represented by Tamarack *(Larix laricina).* The short bluish needles are thickly clustered on stubby spur-like projections, giving the trees a soft, feathery look, and turn a deep golden color in autumn before falling.

The favorite habitat of Tamarack is moist, boggy soil, a preference it shares with Northern White Cedar *(Thuja occidentalis),* which is also known as Arbor Vitae. This tree is instantly recognizable by the flattened branchlets (looking like a freshly pressed herbarium specimen while still on the tree) covered by closely imbricate yellow-green scalelike leaves. The cones sit upright and when open suggest tiny wooden tulips.

Still another conifer, the sight of which often indicates the existence of a cold Sphagnum bog, is Black Spruce *(Picea mariana).* It is a tall, thin tree with short branches unevenly distributed, and by comparison with the full symmetry of most other spruces it gives the deceptive impression of being a loser in the battle with the elements.

Jack Pine *(Pinus banksiana),* on the other hand, grows in dry, barren soil. This tree is unique among eastern pines in having cones that are strongly curved inward toward the branch. It is very abundant in a limited section of Michigan between Saginaw Bay and the Straits of Mackinac, where on the "plains" of just a few counties it holds its own against encroachment by deciduous species. It is helped greatly by frequent fires, which burst open the tightly closed cones and clear the soil for the seedlings. Thick stands of young pines constitute the only nesting habitat acceptable to Kirtland's warbler, and the world's only breeding ground of this rare bird is here in an amazingly small area.

Two deciduous trees are particularly well adapted to compete with conifers in the cooler woods, and evidence of their success can be seen in their enormous ranges. One is Quaking Aspen *(Populus tremuloides),* which comes in quickly as a pioneer following a fire, windstorm, or any other occurrence that admits ample sunlight. Flattened petioles cause its leaves to flutter in the slightest air current, and this accounts for the name. Another opportunistic species is Paper Birch *(Betula papyrifera),* well-known for its chalky white peeling bark.

INDIANA DUNES NATIONAL LAKESHORE

Indiana Dunes National Lakeshore occupies a long strip of land along the southern end of Lake Michigan between Gary and Michigan City. For an introduction to this facility you should first stop at the visitor center on Kemil Road, which can be reached from Valparaiso via Ind. 49 north, then U.S. 20 east. Here an interpretive audiovisual program will be shown on request, and park naturalists will furnish maps, trail pamphlets, and other information.

The half-mile self-guiding Calumet Dune Trail behind the visitor center takes you over old dunes that once stood on the shore of a much larger Lake Michigan but are now covered with an oak-hickory forest, with an abundance of Sassafras *(Sassafras albidum)* and Witch Hazel *(Hamamelis virginiana)* in the understory. A brochure explains the process by which these woods became established on what had been banks of drifting sand. For many, this trail will be most appealing during the spring and early summer months, when such wildflowers as Great and False Solomon's-seal (*Polygonatum canaliculatum* and *Smilacina racemo-*

sa), Intermediate Dogbane *(Apocynum medium),* and Fern-leaved False Foxglove *(Aureolaria pedicularia)* are in evidence.

Other opportunities for botanizing exist in the eastern section of the National Lakeshore (and in Indiana Dunes State Park, which is within its boundaries). These include some rather strenuous hiking trails, a heron rookery, and a disjunct bog; inquiries should be made in advance about their suitability and, in some cases, about permission to enter the areas. In the other direction, at the Bailly-Chellberg complex (reached by turning north from U.S. 20 on Mineral Springs Road, which is 4.5 miles west of Kemil Road) the emphasis is on its historical structures, but the mile-long trail has a good representation of spring wildflowers. The most highly recommended short trail, however, is the West Beach Succession Trail.

West Beach is off County Line Road, which runs north from U.S. 20 about 8 miles west of Mineral Springs Road. The Succession Trail is a 1-mile loop, and to be sure of taking it in the direction followed in the printed brochure you should commence it by entering the arcade near the shore marked Bath House. (Before starting out, however, you may wish to walk down to the intradunal pond opposite the bath house, where insectivorous Horned Bladderwort *(Utricularia cornuta),* Lake Shore Rush (*Juncus balticus* var. *littoralis*), Rose Pink *(Sabatia angularis),* and Bog Lobelia *(Lobelia kalmii)* may be seen. A path that has been laid out to provide access to a "Dune Climb" area makes it possible to get close to this pond.)

The Succession Trail commences over a fore dune behind a popular bathing beach, with the Chicago skyline visible on the horizon. As you ascend over the dune—on wooden stairs and walks to minimize damage by foot traffic—the progression from Marram Grass to Little Bluestem *(Schizachyrium scoparium)* is clearly seen. Extensive areas are blanketed by sprawling Dune Grape *(Vitis riparia).* In back of the dunes, woody plants become more evident: Sand Cherry; Serviceberry (*Amelanchier* sp.); Wafer Ash, or Hop Tree *(Ptelea trifoliata);* a scattering of Red Cedar *(Juniperus virginiana);* Common Juniper (*J. communis* var. *depressa*); and the ubiquitous Eastern Cottonwood. Perhaps the showiest flowers are those of the orange-yellow Hoary Puccoon *(Lithospermum canescens),* while the most inconspicuous may be the delicate white blossoms of Sand Cress *(Arabis lyrata),* whose distinctive basal leaves have a disconcerting habit of vanishing when the plants come into flower. Also here, at the southern extreme of its range, is a grove of Jack Pine, accompanied by another boreal plant, the evergreen Bearberry, or Kinnikinick *(Arctostaphylos uva-ursi).*

Skirting another pond, the trail climbs to the top of an ancient dune now covered by a forest of Black Oak *(Quercus velutina).* Among the spring flowers found here are Canada Mayflower *(Maianthemum canadense),* Starry Solomon's-plume *(Smilacina stellata),* and Columbine *(Aquilegia canadensis).* Emerging from these woods, the trail descends through a colorful mixture of Puccoon, Wild Rose *(Rosa blanda),* and Spiderwort *(Tradescantia ohiensis).*

OHIO SPIDERWORT
Tradescantia ohiensis

THE DOOR PENINSULA

The Door Peninsula is that splinter of eastern Wisconsin that thrusts out into Lake Michigan to form Green Bay. Historically, it was the scene of scores of shipwrecks; in fact, its name is a carryover from the appellation "Door of the Dead," which was given by apprehensive French explorers to the narrow strait between the tip of the peninsula and Washington Island. Today it is a tranquil land of cherry orchards and dairy farms. Although the Door Peninsula owes much of its popularity as a quiet summer resort area to its fairly remote location, it actually is more accessible than it might appear thanks to the availability of car ferry service to nearby Kewaunee from Ludington across the lake in Michigan.

The city of Green Bay is the gateway to the Door Peninsula, and several pleasant hours can be spent there at the Bay Beach Wildlife Sanctuary on East Shore Drive (take Irwin Street north to the end). One-third of its 700 acres is open to the public, and in addition to a nature center there are a half-dozen trails leading through woods, meadows, and marshes.

Whitefish Dunes State Park

Of the several state and county parks located on the Door Peninsula, Whitefish Dunes State Park is outstanding because of its varied terrain. A well-marked access road turns east from S.R. 57 about 8 miles north of Sturgeon Bay; the park entrance is reached at 3.8 miles.

Hikers can choose from a network of trails with an aggregate length of 11 miles. A good way to start would be with the Red Trail, which runs southwest behind the front dunes. Access to the beach from this trail is provided at several points, where the active dunes can be observed in the process of gradually inundating stands of Northern White Cedar and other trees. Only a few non-woody plants, such as Tall Wormwood, are able to eke out an existence here.

On the lee side of these forward dunes, however, the trailsides are lush with vegetation. Thimbleberry *(Rubus parviflorus)* is rampant, with Chokecherry *(Prunus virginiana)* and Mountain Maple *(Acer spicatum)* the principal deciduous trees. Eastern Hemlock *(Tsuga canadensis),* Balsam Fir *(Abies balsamea),* and Canada Yew *(Taxus canadensis)* help to clothe the backside of the dunes. Wild Columbine is a particularly abundant native wildflower, while Leafy Spurge *(Euphorbia esula)* is among the inevitable adventive species.

A right turn at midpoint puts you on a wide, sunlit sandy road with a row of Paper Birches on one side and a field glowing with masses of burnt orange Devil's Paintbrush *(Hieracium aurantiacum)* on the other. The next right leads back to the starting point on a parallel trail only a short distance from the original leg but on a much older dune and with totally different vegetation—a particularly vivid illustration of plant succession. Here the soil is still sandy but has been vastly enriched by humus laid down over a long period of time. The result is an almost pure forest of American Beech *(Fagus grandifolia)* and Sugar Maple *(Acer saccharum).* There are few flowering plants under these—one notable exception being Wild Sarsaparilla *(Aralia nudicaulis)*—but ferns are well represented by Oak Fern *(Gymnocarpium dryopteris),* Rattlesnake Fern *(Botrychium virginianum),* and many clumps of Lady Fern *(Athyrium filix-femina),* among others. Club mosses are also plentiful, including Wolf's Claw, Tree, and Shining Clubmoss (*Lycopodium clavatum, L. obscurum,* and *L. lucidulum*).

Only as you near the end of the trail do wildflowers begin to appear: Rose Twisted-stalk *(Streptopus roseus),* Early Meadow-rue *(Thalictrum dioicum),* and Golden Corydalis *(Corydalis aurea).* Out in the open sun, two less modest plants take over: Cow Parsnip *(Heracleum lanatum)* and Dame's Rocket *(Hesperis matronalis).*

The White Trail is a double loop in the eastern section of the park, where the soil lies in a thin layer over a rock base and the shoreline consists of jagged eroded limestone cliffs instead of drifting sand dunes. The approach road is flanked by Great White Trillium *(Trillium grandiflorum)* in early spring. The trail itself—again through a beech-maple community—has such attractions as

Red-berried Elder *(Sambucus pubens),* White and Red Baneberry (*Actaea pachypoda* and *A. rubra*), and Downy Yellow Violet *(Viola pubescens),* but one gets the impression that the woods are primarily dedicated to Sweet Cicely *(Osmorhiza claytoni),* which grows here by the thousands.

For a change from this comparative monotony, keep to the right all along the trail until you reach a point on the uppermost loop where the trail goes left and an unmarked path leads straight ahead. Take the latter, cross the dirt road, and look for the path to continue a few steps to the left. This leads to a narrow trail that winds along the limestone ledges, following the lakeshore in both directions. In this moist environment you are under Northern White Cedars and Balsam Firs, with bristly Swamp Currant *(Ribes lacustre)* in the shrub layer, and the forest floor is embellished with Bluebead Lily *(Clintonia borealis),* Starflower *(Trientalis borealis),* and Canada Mayflower.

The Ridges Sanctuary

To say that Whitefish Dunes State Park will suffer by comparison after a visit to The Ridges Sanctuary is not to demean one but to extol the other, for The Ridges must qualify as one of the finest natural gardens anywhere. It is to the credit of those responsible for maintaining its integrity that they have not been tempted to introduce plants from elsewhere—a practice that has become all too common, and one that in this case would have been not only deplorable but patently unnecessary.

Operated as a nonprofit entity supported entirely by voluntary donations, The Ridges Sanctuary is open to the public all year. It is located about 12 miles north of Whitefish Dunes, at the north end of Bailey's Harbor just east of S.R. 57 on County Road Q.

The name refers to a series of parallel ridges composed of glacial sand, formed ages ago into submerged sandbars beneath an older and larger Lake Michigan but now richly covered with vegetation. A system of foot trails has been constructed along the ridge crests, with numerous wooden bridges crossing the intervening swales.

Even before one has left the parking area, there is evidence all around that many of the plants to be seen here are those commonly associated with a northern environment: from Thimbleberry and Highbush Cranberry *(Viburnum trilobum)* to Bunchberry *(Cornus canadensis)* and Bearberry. There is such an abundance of Labrador Tea *(Ledum groenlandicum)* that one of the trails has been named for it, but it would be a mistake to concentrate so intently on these attractive shrubs and the fragrant carpets of dainty Twinflower *(Linnaea borealis)* beneath them that you would fail to look for the elusive little Ram's-head Lady's Slipper *(Cypripedium arietinum),* which also makes its home along the Labrador Trail.

Among the spring-flowering herbs that seem to be almost everywhere are

Starflower, Bluebead Lily, and the vivid magenta Fringed Polygala, or Gay Wings *(Polygala paucifolia).* Others with a more limited distribution include One-flowered Wintergreen *(Moneses uniflora),* Purple Avens *(Geum rivale),* and Three-leaved False Solomon's-seal *(Smilacina trifolia).*

The sanctuary is home to an incredible 28 species of native orchids, some plentiful, others maddeningly hard to locate. Yellow Lady's Slippers (*Cypripedium calceolus* var. *pubescens*) are very numerous, and are especially attractive when scattered throughout a moist area green with Horsetails (*Equisetum* sp.). The beautiful Showy Lady's Slipper *(Cypripedium reginae)* is here too, as well as the popular Pink Moccasin Flower *(C. acaule).* Among the less elegant orchids are Early Coralroot *(Corallorhiza trifida),* which is dismissed by many as worthy of only passing notice, and the handsome Striped Coralroot *(C. striata),* which surely deserves better.

The origin of the ridges is most evident where there are long sandy stretches devoid of trees. Here Creeping Juniper and Lake Iris *(Iris lacustris)* account for most of the ground cover, with a sprinkling of Sand Cress providing white accents. Contrasting with these are the wet swales where, at one small bridge, masses of Buckbean *(Menyanthes trifoliata)* thrust their large cloverlike leaves and racemes of densely bearded white flowers out of the dark shallow water. Between these extremes are stands of White and Black Spruce (*Picea glauca*

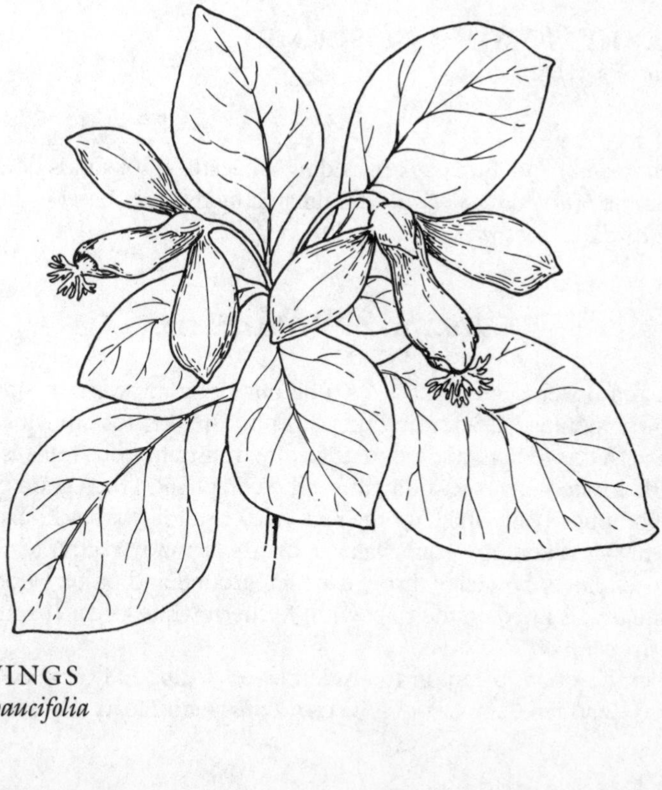

GAY WINGS
Polygala paucifolia

STARRY SOLOMON'S-PLUME
Smilacina stellata

and *P. mariana*) and Tamarack providing a backdrop for fields of Starry Solomon's-plume *(Smilacina stellata)* illuminated by flashes of scarlet Indian Paintbrush *(Castilleja coccinea)*.

LUDINGTON STATE PARK

More than a dozen state parks are strung out along the western side of Michigan's lower peninsula, affording ample opportunities for studying lakeshore vegetation. A convenient and popular facility with numerous trails is Ludington State Park, located north of Ludington at the end of S.R. 116. A good choice here is the Lighthouse Trail, which for the first mile or so follows a well-packed gravel road (which makes for easier walking than soft sandy paths) to Point Sable Lighthouse. The approach is through a campground and past a wooded stretch where the trail is lined on the right with Northern White Cedar shading tall Blue Flags *(Iris schrevei)*.

Out in the open, common roadside flowers include Silverweed *(Potentilla anserina)*, Bladder Campion *(Silene cucubalis)*, and Hoary Alyssum *(Berteroa incana).* Yellow Goatsbeard *(Tragopogon pratensis)* is conspicuous with its yellow composite flowers followed by tawny balls of fluff resembling oversized dandelion seed heads. Much less numerous than these are White Camass *(Zygadenus glaucus)* and Northern Green Orchid *(Platanthera hyperborea).* Shrubby Willows, Sand Cherry, and Red Osier Dogwood *(Cornus stolonifera)* are evident everywhere.

The trail passes parallel to and just behind the front dunes, and in this protected zone such plants as Hoary Puccoon, Sand Cress, and Balsam Ragwort *(Senecio pauperculus)* are able to maintain a purchase in the more stable sand.

Where blowouts have occurred there are ponds backed by dark Jack Pines, some virtually filled with bright yellow Horned Bladderwort and many visited by white-tailed deer in early morning and at dusk. Their sandy rims have creeping Bearberry, Shrubby Cinquefoil *(Potentilla fruticosa),* and clumps of low, straggly Junipers.

Finally, around the base of the lighthouse are White Campion *(Lychnis alba),* with flowers that open in the evening; beautiful red-purple Beach Peas *(Lathyrus japonicus);* and fragrant drifts of Common Milkweed *(Asclepias syriaca).*

McSAUBA RECREATION AREA

Located near the top of Michigan's lower peninsula, Charlevoix is an attractive resort city made especially charming by plantings of pink, white, and blue petunias along every foot of the main street from one end of town to the other.

Since Charlevoix fronts on Lake Michigan only 32 miles from the Beaver Island archipelago, possesses a fine yacht harbor, and has the 12-mile-long Lake Charlevoix at its backdoor, its recreational opportunities are water oriented, but in at least one place—the McSauba Recreation Area—the integrity of the natural environment has been well preserved in close proximity to a popular beach.

To reach the parking lot, turn west from U.S. 31 on Mercer Boulevard for 0.8 mile, then left on Pleasant Street. About 0.3 mile farther, where the street curves to the right, drive straight ahead on the dirt road through the open gate.

A short walk leads through a grove of Paper Birches to the shore of Lake Michigan. Conspicuous along the way are several groups of Red Osier Dogwood, and there is quite a lot of Starry Solomon's-plume, its berries marked with broad dark stripes while still immature. More to be expected is the presence of a number of opportunistic European species with a propensity for quickly moving into disturbed areas: Blueweed, or Viper's Bugloss *(Echium vulgare);* Rough-fruited Cinquefoil *(Potentilla recta);* Cypress Spurge *(Euphorbia cyparissias);* and Hoary Alyssum, for example—plants that possess undeniable visual appeal but whose invasive nature sometimes engenders more resentment than appreciation.

The situation changes abruptly, however, for no sooner do the dunes come into view than you recognize two plants as being unusual. One is the Great Lakes

GREAT LAKES THISTLE
Cirsium pitcheri

Thistle *(Cirsium pitcheri),* listed as threatened in the state. The cream-colored flower heads and the finely cut foliage, which appears gray because of the white tomentum, are distinctive. The other, also a composite, is Lake Huron Tansy *(Tanacetum huronense),* a plant more widespread than the name would suggest but much rarer than the Common Tansy *(T. vulgare)* of Old World origin, which has escaped from cultivation and established itself over a broad range. The yellow heads of the native species are fewer but larger, and frequently include short ray flowers. Together with the dark green fernlike foliage, they stand out in strong contrast against the background of dazzling white sand.

For a change, you will want to return over the network of trails that weave through the woods. This is a forest rich in northern gymnosperms, among them Balsam Fir, Northern White Cedar, and American Yew. In the spring it is replete with flowering plants, including Large-flowered Trillium, Bluebead Lily, Twisted-stalk, Canada Mayflower, Wild Columbine, Wild Sarsaparilla, Bunchberry, and Maple-leaved Viburnum *(Viburnum acerifolium).* Later there is a great abundance of Helleborine *(Epipactis helleborine),* an alien orchid always interesting for its variations in color, accompanied by Enchanter's Nightshade *(Circaea quadrisulcata),* Common St. John's-wort (*Hypericum perforatum),* and Harebell *(Campanula rotundifolia*).

WILDERNESS STATE PARK

Farther north, at the extreme northwestern tip of the peninsula, is Michigan's Wilderness State Park, reported to be one of the best in the system for wildflowers. By taking exit 337 west from I-75, the park entrance is reached in about 8 miles.

Several trails are available here, and you may wish to consult blooming schedules at the park headquarters before starting out. Or you can turn left (south) on the road opposite the camper registration area, park in the designated space, and walk past the barricade (disregarding the right fork) on Swamp Line Road.

There is no shade along this path at midday, and some kind of head covering might be advisable, but this is conducive to the multitudes of sun-loving plants that adorn the roadsides. Among the most prolific of these are Twinflower, Bunchberry, Blue-eyed Grass *(Sisyrinchium albidum),* Bastard Toadflax *(Comandra richardsiana),* pale pink Common Fleabane *(Erigeron philadelphicus),* white Star-thistle *(Centaurea diffusa),* and Spurred Gentian *(Halenia deflexa).* All except the last add much to the beauty of the surroundings; when compared with some of the striking wildflowers in the Gentian family, those of *Halenia* are always disappointing. In early summer these are joined, in lesser numbers, by Pink Pyrola *(Pyrola asarifolia),* Marsh Vetchling *(Lathyrus palustris),* gleaming white Canada Anemone *(Anemone canadensis),* and Northern Green Orchid *(Platanthera hyperborea).* Open moist spots are inhabited by False Asphodel (*Tofieldia racemosa* var. *glutinosa*) and Marsh Bellflower *(Campanula aparinoides).*

To the left of the road, past dark green Buffalo Berry bushes *(Shepherdia canadensis),* glimpses may be had of Goose Pond, with large areas covered by the floating leaves of Spatterdock *(Nuphar variegatum),* and a little farther along, of a beaver lodge in the distance. On the other side, a wide grassy opening has a profusion of yellow flowers: Tickseed *(Coreopsis lanceolata)* and Silvery Cinquefoil *(Potentilla argentea),* followed by St. John's-worts (*Hypericum* spp.). Then, after bisecting a Cattail marsh, the trail enters the woods through a stand of Tamarack over Royal Fern *(Osmunda regalis),* with Yellow Lady's Slippers lighting up the deep shade off to the sides.

A road branching off to the left through a Black Spruce swamp is worth exploring for a short distance, at least. There are Labrador Tea and beds of Lake Iris along the roadsides and, in standing water, Marsh Marigold *(Caltha palustris)* and Tufted Loosestrife *(Lysimachia thyrsifolia).* Picking your way through these wet woods by stepping on sphagnum-covered hummocks, you may discover colonies of Blunt-leaved Orchid *(Platanthera obtusata)* and the extremely tiny Naked Miterwort *(Mitella nuda).*

As you return to the main road and continue to walk south, look for the exquisite rose and white blooms of Showy Lady's Slipper *(Cypripedium*

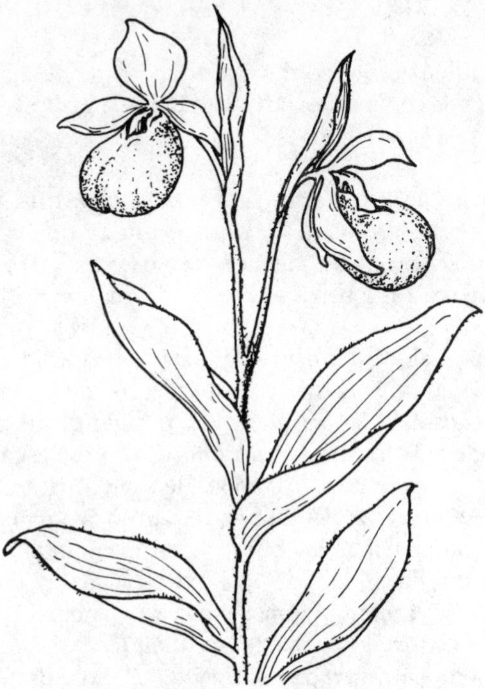

SHOWY LADY'S SLIPPER
Cypripedium reginae

reginae). The plants may be seen very close to the edges, and sometimes even in the middle of the roadway, and because of their large size would be hard to overlook even when not in flower. Other interesting plants with a predilection for wetness will be spotted along here: the tall, geometrically precise spires of Water Horsetail *(Equisetum fluviatile),* dense stands of Cattails *(Typha latifolia),* and Buckbean.

At one point both Red and White Baneberry have been observed, separated by just a few yards. All were in fruit, and one specimen with white berries was clearly *A. rubra* forma *neglecta* since it lacked the thick, pink pedicels of *A. pachypoda.*

Intermittently there are drier areas, notably one where the road curves and where White and Red Pines (*Pinus strobus* and *P. resinosa*) have joined the other conifers. Although much of this section is covered by Bracken *(Pteridium aquilinum)* and Wintergreen *(Gaultheria procumbens),* there is plenty of room left for a sizable population of Pink Moccasin Flowers.

Beyond this piny wood there are more wet areas—and more Showy Lady's Slippers. Approximately 2 miles from the start, the road ends at the junction with Sturgeon Bay Trail on the right and South Boundary Trail on the left.

GRASS BAY PRESERVE

It would be only natural to expect that The Nature Conservancy's Grass Bay Preserve, with its diversity of habitats, should produce an impressive list of plant species, but the reality is no less than astounding. The present count exceeds 300 vascular plants, including 25 orchids and several rarities.

Situated on the Michigan shore of Lake Huron about 6 miles east of the junction of U.S. 23 and S.R. 27 in downtown Cheboygan, the entrance is marked by three posts at the foot of the bank on the north side of U.S. 23. Almost immediately there is a challenge, for intermixed with Buttercups at the very start of the trail are Yellow Avens *(Geum aleppicum);* the two plants can hardly be considered similar, but they correspond just closely enough in flower size, color, and general aspect for the differences to be easily overlooked.

Passing between ranks of Bush Honeysuckle *(Diervilla lonicera)* and white-blossomed Thimbleberry, the main trail descends gradually through a stand of Red Maple *(Acer rubrum)* brightened by Paper Birches. Wintergreen is very abundant, but the related Creeping Snowberry *(Gaultheria hispidula)* is less so. A close look along the edges of the path will detect fragile Daisyleaf Grape Ferns *(Botrychium matricariifolium)* standing less than 4 inches high.

A short detour onto an old road to the right leads into a White Pine and Eastern Hemlock forest with jade green pads of Pincushion Moss *(Leucobryum glaucum)* underfoot. Here may be seen both Lesser and Menzies' Rattlesnake Plantains (*Goodyera repens* and *G. oblongifolia*).

Returning to the main trail, you descend through hemlocks to a log bridge over a small stream. Just before you reach the stream, look in a low place on the right for a large colony of Naked Miterwort, diminutive plants with flowers of an incredibly delicate design, such as might be spun by the tiniest of spiders. In and near the water there is a predictable assortment of Northern White Cedar, Marsh Marigold, Royal Fern, Swamp Candles *(Lysimachia terrestris),* and Purple Fringed Orchids *(Platanthera psycodes).*

Continuing straight ahead over a second log crossing, you soon reach the shore of Lake Huron, where such carnivorous plants as Pitcher Plants *(Sarracenia purpurea),* Round-leaved Sundew *(Drosera rotundifolia),* and Horned Bladderwort share the boggy habitat with Lake Iris and Goldthread *(Coptis groenlandica)*, illustrated on the next page.

To more easily follow the shoreline, retrace your steps a short distance and turn left on the first road, which runs between Lake Huron and a small interior body of water known as Grass Lake. Along this route, there are open places densely covered with Twinflower, Bearberry, and Trailing Arbutus *(Epigaea repens),* punctuated by Harebell, Fringed Polygala, False Asphodel, and Slender Ladies' Tresses (*Spiranthes lacera* var. *gracilis*). Along the banks bordering the woods there is an unusual concentration of plants belonging to the Pyrolaceae: Pipsissewa *(Chimaphila umbellata),* Round-leaved Pyrola (*Pyrola rotundifolia* var. *americana*), and Greenish Pyrola *(P. virens),* all quite numerous, as

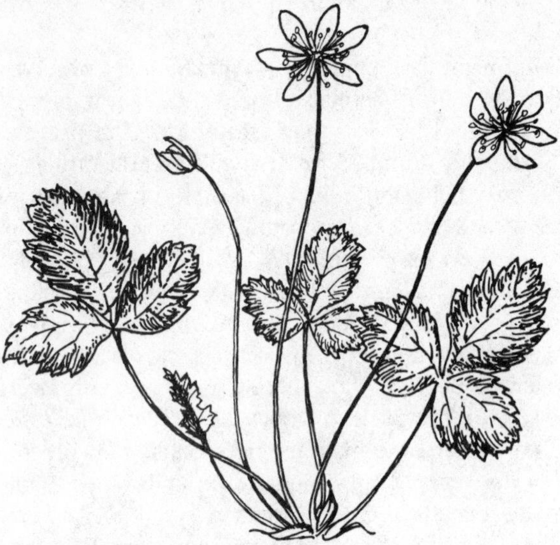

GOLDTHREAD
Coptis groenlandica

well as Indian Pipe *(Monotropa uniflora),* Pinesap *(M. hypopithys),* and Pinedrops *(Pterospora andromedea).*

Shortly the path veers out to Lake Huron, this time with a good view of Bois Blanc Island and its lighthouse. Here are more Pitcher Plants, as well as Indian Paintbrush, Sweet Gale *(Myrica gale),* and Wood Lilies *(Lilium philadelphicum),* and as the trail climbs over the dunes there are Wild Roses, Smooth Hawkweed *(Hieracium florentinum),* and Lake Huron Tansy. At many points along this road it is possible to walk through the cedars on your right to the reedy banks of Grass Lake, where the principal flowering plants are Tall Blue Flag and Tufted Loosestrife.

19

The Tallgrass Prairies

FOR CENTURIES the intricate forces of nature kept the grasslands of the nation's heartland in exquisite balance. East of the country's central axis, in a belt extending from Canada down into Texas, lay the true prairies. There the land was favored by a climate that conserved moisture, and responded with 8-foot grass and hundreds of different wildflowers. Toward the western mountains, as more arid conditions prevailed, there were progressively shorter species of grasses and fewer kinds of forbs.

Bison by the millions roamed the tall grass and multitudes of pronghorn antelope populated the western plains. Fires started by lightning or, less frequently, by the Indians, swept across vast expanses. But grazing and burning were more beneficial than harmful. They pruned back excessive growth, stopped invading woody plants in their tracks, removed suffocating accumulations of dry litter, and released valuable nutrients to the soil. The indigenous plants had their root systems safely deep in the rich loam, and the grasses in particular, with their unique ability to continue growing from the base, suffered little more than we do from a haircut.

All this was to change drastically after the Civil War as homesteaders spread out from the East and, finding what they perceived to be an infinite abundance of fertile land, set about bending it to their will. The worst to suffer was the tallgrass prairie. Breaking up the tough matted sod was a fierce challenge, but it yielded to the steel plow—one of many products of the industrial revolution to exert a profound influence upon the development of the nation. The remarkably diverse and resilient native plant communities were replaced by single crops, such as corn or wheat, which took much from the soil but gave back little. The fine topsoil became vulnerable to erosion, first by rain and then by wind, and the inevitable climax was the Dust Bowl of the 1930s.

There was a time when the tallgrass prairie ecosystem reached into more than a dozen states and covered a total of 400,000 square miles, but the little that remains today equates to something like 1 percent of its original extent. Some sites are pathetically small and have survived only because of the fortuitous protection accorded by their locations—in pioneer cemeteries, along railroad sidings, in roadside ditches, or among outcroppings of rock. Fortunately, the prospects of restoring larger tracts of prairie are improving as a result of a growing awareness of what we have lost and of the need for taking protective measures while there is still time.

HOOSIER PRAIRIE NATURE PRESERVE

Opportunities for preserving relatively large expanses of tallgrass prairie are more likely to exist in sparsely populated states, but it is encouraging to see that conservation groups are also concentrating on examples that are conveniently accessible to the millions who live in big cities like Chicago.

One such site is Hoosier Prairie. Located near Griffith in the northwestern corner of Indiana, it is the largest remnant of prairie in that state and is managed by its Department of Natural Resources. It is most easily reached by taking U.S. 41 south from I-80 for approximately 3.5 miles, then going east on Main Street; the parking lot is on the right after crossing Kennedy Avenue.

A mile-long trail with a double-loop configuration tracks through a portion of the preserve and provides a generous sampling of the extreme diversity of plant species for which the tract as a whole is noted. Ostensibly level, the topography of Hoosier Prairie actually contains subtle variations in elevation and, consequently, in moisture conditions. There are dry sand rises with Black Oak *(Quercus velutina)* savannas—an outstanding feature—as well as mesic prairies, sedge meadows, and marshes, each with its complement of herbaceous species.

Part of the land traversed by the Hoosier Prairie Trail was farmed as recently as 1974. Upon cessation of this activity it was quickly taken over by alien species, but the native prairie vegetation is gradually and successfully recapturing the plot.

One striking feature is the abundance of ferns. The wetter areas have large clumps of Royal Fern *(Osmunda regalis)* or light green beds of Sensitive Fern *(Onoclea sensibilis).* Where it is drier there are fields of Bracken *(Pteridium aquilinum)* overtopped by blue Spiderwort *(Tradescantia ohiensis)* and white Wild Quinine *(Parthenium integrifolium),* with that pretender, the pungently aromatic Sweetfern *(Comptonia peregrina),* in close attendance.

Several dainty wildflowers venture out to the trail edges: Bluets *(Houstonia caerulea),* Cynthia *(Krigia biflora),* and Pale Lobelia *(Lobelia spicata)* to mention a few, but more robust plants seem to be the rule. Some of these are Rattlesnake Master *(Eryngium yuccifolium),* White Wild Indigo *(Baptisia leucantha),* Prairie

BLACK-EYED SUSAN
Rudbeckia hirta

Alumroot *(Heuchera richardsonii),* Goat's Rue *(Tephrosia virginiana),* and Swamp Saxifrage *(Saxifraga pensylvanica).* Composites are, of course, well represented in summer by such species as Black-eyed Susan *(Rudbeckia hirta),* Yarrow *(Achillea millefolium),* Prairie Coreopsis *(Coreopsis palmata),* Boneset *(Eupatorium perfoliatum),* and Blazing Stars (*Liatris* spp.).

The second loop puts on an especially attractive display in June with a mixture of Spiderwort, False Solomon's-seal *(Smilacina racemosa),* Prairie Phlox *(Phlox pilosa),* and dazzling masses of foamy white Northern Bedstraw *(Galium boreale).*

GOOSE LAKE PRAIRIE NATURE PRESERVE

In some ways, Goose Lake Prairie is Illinois' counterpart of Hoosier Prairie. It is the biggest tract of prairie remaining in the state, is near the metropolitan area, and has available for public use a section of the preserve complete with a twin-loop interpretive trail. Perhaps the most obvious dissimilarity noticed by the average visitor is that, instead of oak savannas as at Hoosier Prairie, Goose Lake Prairie has broad expanses that were infiltrated only in fairly recent times by a few trees.

The visitor center is reached from I-80 by turning south on S.R. 47 through Morris for 3.8 miles, left on Pine Bluff Road for 5 miles, and left again on Jugtown Road. The trail, which is 1.5 miles long, starts here.

Near the trailhead, one is struck by the number of "weedy" plants that have taken over in the wake of soil disturbance. There are not only multitudes of exotics like Yellow Sweet Clover *(Melilotus officinalis),* Alsike *(Trifolium hybridum)* and other true clovers, fruit-scented Pineapple Weed *(Matricaria matricarioides),* and golden yellow Wild Parsnip *(Pastinaca sativa)* but also some North American species that are hardly less overwhelming: Ragweed *(Ambrosia artemisiifolia),* Buckhorn *(Plantago aristata),* Horse Nettle *(Solanum carolinense),* and Common Milkweed *(Asclepias syriaca)* among them.

As you progress along the path, though, it is soon evident that the prairie flora is recovering as you begin to see stands of native grasses and more and more forbs like White Wild Indigo, Rattlesnake Master, Phlox, and Coreopsis—and, almost hidden by the higher growth—Wild Roses *(Rosa carolina)* and Sundrops *(Oenothera pilosella).* Management practices that are assisting in the resurgence of this vegetation include selective burning. This simulates, on a much smaller scale, the effects of naturally caused fires in the past and checks the invasion of the grasslands by Hawthorns (*Crataegus* spp.) and other trees.

At the top of the upper loop, a floating bridge crosses a pothole pond rimmed by Cattails backed up by masses of Spiderwort. Aquatic plants that can be examined here at close range include Arrowhead (*Sagittaria latifolia* and *S. graminea*) and Great Bur Reed *(Sparganium eurycarpum).*

Above all, Goose Lake is a place to study prairie grasses, and the booklet describing the trail is an invaluable aid to learning at firsthand about such components as Big and Little Bluestem (*Andropogon gerardi* and *Schizachyrium scoparium*), Switchgrass *(Panicum virgatum),* Indiangrass *(Sorghastrum nutans),* Prairie Dropseed *(Sporobolus heterolepis),* Sloughgrass *(Spartina pectinata),* and Bluejoint *(Calamagrostis canadensis).*

ILLINOIS BEACH STATE PARK

Occupying a 6.5-mile strip of the Lake Michigan shore near the town of Zion, just south of the Wisconsin border, is Illinois Beach State Park. The northern unit, which is entered via 17th Street eastward from Sheridan Road, is dedicated to day use and has a network of trails totaling about 5 miles. These afford an opportunity to walk through several areas of sand prairie on and between the stable ridges behind the front dunes. Here in the coarse sandy soil are to be found Black-eyed Susan, Coreopsis, Prairie Phlox, Blazing Stars, Prairie Alumroot, and the lovely pink Shooting Star *(Dodecatheon meadia).* Where more moisture is present there are Canada Anemone *(Anemone canadensis),* Rock Sandwort *(Arenaria stricta),* and Blue Flag *(Iris schrevei).*

In places where the path skirts the backs of active dunes, Balsam Ragwort *(Senecio pauperculus),* Hoary Puccoon *(Lithospermum canescens),* and Silverweed *(Potentilla anserina)* are occasionally joined by Wild Four-o'clock *(Oxybaphus nyctagineus),* an odd plant with little flowers arising from cuplike

CANADA ANEMONE
Anemone canadensis

involucres in pathetic imitation of its glamorous relative, *Bougainvillea.* Unfortunately, the picture here is somewhat distorted by an assortment of exotic pines and horticultural forms of honeysuckle, spirea, and other shrubs—souvenirs of plantings made prior to acquisition by the state and presumably intended to "improve upon" the indigenous vegetation.

The main entrance to the park on Wadsworth Avenue leads to the southern unit, where another series of foot trails affords marked contrast. An interesting circular walk of less than 2 miles may be had by taking the Dead River Trail and returning via the Oak Ridge Trail.

On the outward leg there is a Black Oak savanna on the left and the marsh-bordered river on the right. (For much of the year the Dead River is actually a long pond, its outlet blocked by a sand bar thrown up by storm waves.) A particularly striking flower that occurs near the beginning in an open grassy area is Downy Painted Cup *(Castilleja sessiliflora),* with prominent yellow corollas. Other plants along the way include Canada Anemone; the smaller, greenish-flowered Thimbleweed *(Anemone cylindrica);* Marsh Vetchling *(Lathyrus palustris);* Canada Mayflower; and Blue Flag.

After crossing a small bridge you will come to a patch of Shrubby Cinquefoil *(Potentilla fruticosa)* and a spread of Starry Solomon's-plume *(Smilacina stellata).* Turn left at each succeeding junction until you get to another wood chip (not gravel) trail; this is Oak Ridge Trail, and for its entire length it passes through a relatively open but pleasantly shaded savanna with Black Oaks and Chokecherries *(Prunus virginiana)* the most numerous tree species.

The return along this sandy ridge is pure delight in early June when there are masses of mixed blooms at every turn. The luminous colors of Coreopsis, Prairie Phlox, and Puccoon subdued by the misty white puffs of Redroot *(Ceanothus ovatus)* are an unforgettable sight.

SCUPPERNONG SPRINGS NATURE AREA

These days prairies are where you find them, and there is no reason why a specimen tract should not occur in the midst of what appears to be a forest trail less than 40 miles from downtown Milwaukee, and that is exactly the case at Scuppernong Springs Nature Area.

To find this spot, drive north from the town of Eagle, Wisc., on S.R. 67, and when you reach the intersection with County Road ZZ go left for one-half mile. (If you miss this turn you will come to some longer hiking trails, which also bear the name Scuppernong.)

Starting out in a counterclockwise direction on the wooded 1.5-mile elliptical trail, you pass the ruins of an old hotel and before long emerge onto a sunlit sandy ridge. Except for some Wild Rose bushes, the only woody plants are some small Black Oaks—an indication that without interference (which may be assumed) this little segment of prairie will eventually succeed to a sand savanna community. Plants like Tall Wormwood *(Artemisia caudata),* Hoary Alyssum *(Berteroa incana),* Ground Cherry *(Physalis virginiana),* and Bush Clovers (*Lespedeza* spp.) are thriving in the absence of a canopy, but many of the more attractive flowering species may be seen as well: Spiderwort, Puccoon, Black-eyed Susan, Goat's Rue, Gayfeather *(Liatris pycnostachya),* and Whorled Milkweed *(Asclepias verticillata)* among them.

There is, of course, much more to the trail. At the apex of the loop there is Scuppernong Spring itself, bubbling up from the sand and forming a large pond behind a dam. In spring, especially, there are flowers virtually everywhere along the path: the pink-striped bells of Spreading Dogbane *(Apocynum androsaemifolium),* a stunning combination of Cynthia and Northern Bedstraw, Yellow Pimpernel *(Taenidia integerrima),* the creamy white doilies of Elderberry *(Sambucus canadensis),* Marsh Marigolds, and Wild Geraniums.

This is an interpretive nature trail, and its 40 explanatory signs contain illustrations and provide more information than usual. Of particular value are those pertaining to specific trees, such as Shagbark Hickory *(Carya ovata),* Quaking Aspen *(Populus tremuloides),* and Bur Oak *(Quercus macrocarpa).*

20

The American Rockies

THE ROCKY MOUNTAINS system begins with the Brooks Range, above the Arctic Circle in Alaska, and ends a mere couple of hundred miles short of the Mexican border. Along this grand sweep, which makes it the longest chain on the continent, there are many approaches. Each affords a different aspect of these lofty ridges and peaks, but the dramatic impact of their first sighting is perhaps most appreciated by westbound travelers as they head toward central Colorado after having driven across the homogenized landscape of Kansas or Nebraska.

When the Rockies first come into view they appear to be just a few miles away, have the clean, sharp look of having been recently formed, and seem to rise abruptly like stage scenery from a perfectly flat foreground—but these impressions are all illusory. The atmospheric haze that distorts spatial perspective and that many of us have come to accept as normal is virtually absent from the dry air of the plains, and there are no intervening land forms to lend scale to the distance. If we carry in our minds the image of old, time-rounded mountains like the Appalachians, then certainly the Rockies are young, but only by comparison; already they have lost thousands of feet of their original height through eons of erosion. And the plain above which they loom, far from being level, slopes upward steadily from east to west until the *base* of the mountains is a mile above sea level—the result of both the uplift that created the range in the first place and later deposition of sediment washed down from its flanks.

Starting from such a relatively high elevation, it is obvious that we will traverse very few bioclimatic or life zones before reaching heights where plants are barely able to survive at all. In fact, the national parks that we will be visiting—Rocky Mountain, Grand Teton, Yellowstone, and Glacier—contain only three of these belts. The altitudes at which they occur are irregular because of variations in a number of environmental factors, but it can be stated as a general rule that the montane zone will be found between 6,000 and 9,000 feet and the subalpine zone from 9,000 feet up to timberline, which in the Rockies averages around 11,500 feet. Above that, of course, is the alpine zone.

A reliable indicator of the montane zone is Ponderosa Pine *(Pinus ponderosa).* This is sometimes accompanied by Douglas Fir *(Pseudotsuga menziesii),* but more characteristically it covers large areas with open, even-aged forests. Two-needled Lodgepole Pines *(Pinus contorta)* form more dense stands, growing so close together as to limit the foliage to the very tops, which accounts for the common name. It is the most widespread tree in Yellowstone and Grand Teton national parks. A pioneer, it is quick to repopulate a burn but unable to tolerate being shaded out by other kinds of trees.

The light-barked Quaking Aspen *(Populus tremuloides)* is often an indicator of disturbed areas, and is especially eye catching when its foliage takes on a rich golden color in the fall. Other noteworthy species are the familiar Blue Spruce *(Picea pungens),* which grows along mountain streams, and Rocky Mountain Juniper *(Juniperus scopulorum),* which favors low, south-facing slopes. Two shrubs often found together in sunny situations (and sharing the characteristic of wedge-shaped, three-toothed leaves although not at all related) are Antelope Brush *(Purshia tridentata)* and Sagebrush *(Artemisia tridentata).*

Moving up into the subalpine belt we find the lushest vegetation in the Rockies, owing to abundant snowfall and the protection of thick forestation, principally by Engelmann Spruce *(Picea engelmannii)* and Subalpine Fir *(Abies lasiocarpa).* Conspicuous shrubs are Greene's Mountain Ash *(Sorbus scopulina)* and Twinberry Honeysuckle *(Lonicera involucrata),* but the most attractive feature of this zone is the rich variety of herbaceous wildflowers in its meadows.

The upper part of the subalpine zone is transitional, and the trees gradually become dwarfed and twisted by the elements, some having become "banner trees" where the prevailing winds prevent branches from growing except on the lee side, others having been beaten down into matted *krummholz.* In this hostile environment those Spruces and Firs that are able to cling to life are joined by a few other low-growing trees, notably Birches and Willows.

Eventually we come to the true alpine zone, where even the toughest of these woody species has had to give up the fight but where, strange as it may seem, the battle has been joined with resounding success by a host of pretty, colorful little flowering herbs. All perennials, they burst into bloom soon after the snows have melted—and sometimes before—in order that they may set and disperse their seed before the return of winter, which up here is a scant 2 months away.

ROCKY MOUNTAIN NATIONAL PARK

The campsite should not have been all that great. It was only 7 miles from the town of Estes Park and barely inside the boundary of the 412-square-mile Rocky Mountain National Park. We had left Denver right after the Fourth of July in

100-degree heat, and were lucky to have found any spot at all for there seemed to be more than enough takers for the Moraine Park Campground's 260 campsites. As it turned out, though, it was as perfect an introduction to the Colorado Rockies as could have been wished.

One thing that helped was that although it was close to the road the site was insulated by a low rise from the sight and sounds of other campers and traffic. The elevation was 8,200 feet, and we were near the bottom of a south-facing lateral moraine. A typical parklike Ponderosa Pine forest was at our backs, and a short distance in front of the camp this gave way to a rocky, sunlit slope with scattered low shrubs and herbs. At its foot Moraine Park itself began—not the usual U-shaped glacial valley but, thanks to its location just behind a terminal moraine, a wide, flat expanse of meadows and marshes laced with creeks and studded with ponds, and stretching far across to the other side where a 600-foot-high band of dark green defined the opposite lateral moraine. Above this were craggy ridges, and over their rims the tips of higher peaks crowned with snow gleamed against an incredibly blue sky.

The animal life was a constant delight. Daybreak brought mountain bluebirds and an occasional deer. Broad-tailed hummingbirds zoomed in on anything red, making a special effort to extract nectar from a plastic-topped thermos. A bag of nuts absent-mindedly abandoned on the table drew a parade of admirers: chipmunks, ground squirrels, Steller's jays, and magpies.

But it was the wealth and diversity of the plant life at our very "door" that was so irresistible. Every few steps down the gentle incline of the sunny bank brought something new: the pale yellow blossoms of Antelope Brush; stands of the unmistakable One-sided Penstemon *(Penstemon unilateralis),* a tall plant with purple-blue flowers; the clustered blooms of Winged Buckwheat *(Eriogonum alatum);* and mounds of white-flowered Mountain Balm *(Ceanothus velutinus).* Contrary to their name, the Evening Primroses opened their snow white cups in the morning and by evening withered to a lovely pink. There were two species, *Oenothera nuttallii* and *O. coronopifolia,* the latter with finely dissected, fernlike foliage.

By a path at the edge of the meadow there grew handsome Pink Milkweeds *(Asclepias speciosa)* and White Geraniums *(Geranium richardsonii),* and at streamside a few stately Monkshoods *(Aconitum columbianum).* It was in the wet meadows, however, that multitudes of jewellike wildflowers were to be seen. Shooting Stars *(Dodecatheon pulchellum)* and orange-spotted Cinquefoil *(Potentilla pulcherrima)* were two with well-deserved specific names. Lupines included *Lupinus parviflorus,* whose undersized flowers were compensated for by their numbers. Indian Paintbrush *(Castilleja linariaefolia)* stood flame tipped among the grasses in the company of a yellow Owl Clover *(Orthocarpus luteus).* Several species of Lousewort were there, but the most fascinating by far was *Pedicularis groenlandica,* each flower in its dense racemes bearing a startling resemblance to the head of a little bright pink elephant complete with a long curved trunk.

COLORADO COLUMBINE
Aquilegia caerulea

All this was not for our exclusive benefit, of course, but could be enjoyed by all who camped at, or passed through, Moraine Park. The same is true of the many fine trails in this section of the national park, including a quarter-mile nature trail.

Just west of the campground is the trailhead for hiking to Cub Lake, a serene body of water rimmed by conifers. Some interesting plants along this walk are Wild Four-o'clock *(Oxybaphus hirsutus);* Twisted-stalk *(Streptopus amplexifolius);* Fairybells *(Disporum trachycarpum),* which bears whitish flowers early followed by scarlet fruits; Tasselflower *(Brickellia grandiflora),* a strange composite with nodding rayless flower heads; and the striped pink Sticky Geranium *(Geranium viscosissimum).* Growing on the wet shores of Cub Lake and blooming somewhat later is Marsh Gentian *(Gentiana strictiflora),* an erect species bearing numerous dull bluish flowers.

A paved road south from Moraine Park takes you to Bear Lake. Although this appears hemmed in by steep rock walls, you can walk completely around it on a half-mile nature trail. Here you can get an excellent sampling of subalpine terrain and its characteristic trees, including not only Engelmann Spruce and Subalpine Fir but Limber Pine *(Pinus flexilis)* and Rocky Mountain Maple *(Acer glabrum)* as well.

From here one of the most scenic trails in the park leads westward in succession to Nymph, Dream, and Emerald lakes, each more beautiful than the last. Again, a handful of wildflower species will have to suffice as an indication of what is in store. First of all, there is the elegant blue and white Colorado Columbine *(Aquilegia caerulea),* a superb choice for the state flower. There are Penstemons of many kinds, but most unusual is the claret-colored *Penstemon whippleanus.* Another mimicking Lousewort is the white *Pedicularis racemosa,* called Parrots-beak. Among the shrubs are Bog Laurel *(Kalmia polifolia)* and Waxflower *(Jamesia americana).* At Emerald Lake there is a temptation to scramble up the talus slope, not only for the enhanced view but to search out other plants such as Honey Polemonium *(Polemonium brandegei)* and Greenleaf Chiming Bells *(Mertensia viridis).*

For a greater measure of isolation without having to hike into the backcountry, try the Wild Basin, which is located in the southeastern corner of the park and offers a selection of trails. It is reached by driving south from Estes Park or west from Lyons on S.R. 7, then turning west on one of several short roads between Meeker Park and Allenspark to a trailhead.

The Calypso Cascades Trail from Allenspark is of modest length but passes through luxuriant subalpine country with a great variety of trees and wildflowers. A number of orchids will be noted, including Twayblades (*Listera* spp.), Northern Green Orchid *(Platanthera hyperborea),* Coralroots (*Corallorhiza maculata* and *C. trifida*), and—a particularly gratifying find—the dainty Fairy Slipper *(Calypso bulbosa),* from which the trail takes its name. Three Pyrolas

FAIRY SLIPPER
Calypso bulbosa

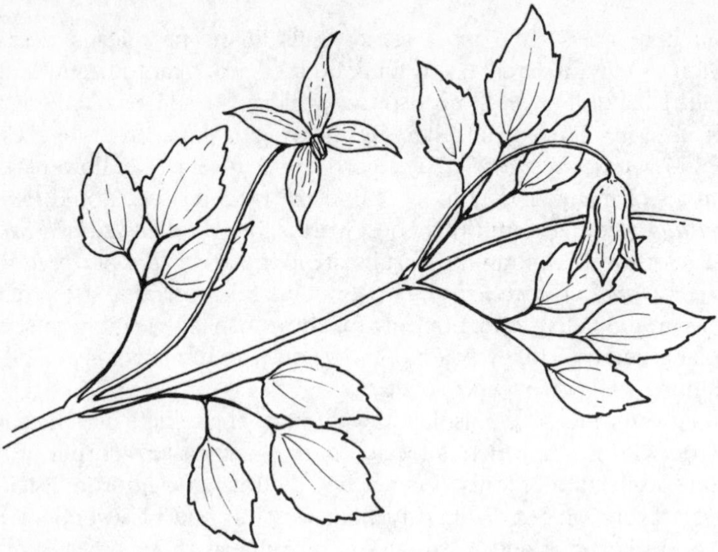

ROCK CLEMATIS
Clematis columbiana

grow here: Pink, Greenish, and Least (*Pyrola asarifolia, P. virens,* and *P. minor*). An unusual vine is the semiwoody Rock Clematis *(Clematis columbiana)* with large blue flowers.

Trail Ridge Road

The few walks just described account for an infinitesimal fraction of the trail mileage available within Rocky Mountain National Park (which aggregates more than 350 miles). Yet the park's most outstanding artery is not a hiking trail but a fine highway—the highest paved road in the country.

Trail Ridge Road (U.S. 34) winds across the park from east to west at altitudes up to 12,183 feet, providing a rare opportunity to observe the alpine ecosystem at close range. The route that it follows and the name it has been given both derive from a high country trail laid out by Ute and Arapaho Indians in ancient times. Normally it is open from the end of May to mid-October, but this is dictated by snow conditions. For alpine flora, the first half of July is probably the best time of all.

A 40-mile section serves as a self-guiding auto tour, beginning at Deer Ridge Junction (with U.S. 36) and ending at Grand Lake. Describing the many features to be seen from these breathtaking heights must be left to the printed brochure, but there are particular places along the route that should be pointed out to those whose interest is botanically oriented.

In the vicinity of Hidden Valley, for example, a path goes up through a subalpine forest beside a small stream with Parry Primrose *(Primula parryi),* Subalpine Jacob's Ladder *(Polemonium delicatum),* and Narcissus-flowered Anemone *(Anemone narcissiflora)* among the showy flowers. Across the road on an open rocky bank there are Colorado Columbine and a few tundra plants like Fairy Primrose *(Primula angustifolia)* and Alplily *(Lloydia serotina).*

Forest Canyon Overlook has a short pathway where other alpine species may be seen, and at Rock Cut there is the interpretive Roger Toll Memorial Trail (paved to protect the fragile plant life). This leads over a boulder-strewn landscape that would be desolate if it were not for the brilliantly colored lichens and the flowers that transform it each year into a natural rock garden. Of necessity, the flowering plants in this treeless, wind-whipped near-arctic environment share diminutive size and reduced foliage, but there the resemblances cease. In form, they vary from round-leaved Mountain Sorrel *(Oxyria digyna),* with sober racemes of dull red, to Alpine Avens *(Geum rossii),* profusely set with bright yellow blossoms, and from low clusters of deep pink Fairy Primroses to the little white "drumsticks" of Snowball Saxifrage *(Saxifraga rhomboidea).* Everywhere you see miniature relatives of well-known garden flowers: Cushions of Tufted Phlox *(Phlox condensata)* with flowers so tightly packed that foliage appears to be totally absent, Greenleaf Chiming Bells that recall the taller Virginia Bluebells, Alpine Wallflower *(Erysimium nivale)* with yellow flowers on inch-long stems, and pink Moss Campion *(Silene acaulis)* with

PARRY PRIMROSE
Primula parryi

DWARF CLOVER
Trifolium nanum

no flower stalks at all. Similar to this last is Dwarf Clover *(Trifolium nanum),* in which the heads are virtually stemless and have been reduced to just a few, or in some cases a single floret each. There is even a tiny Forget-me-not *(Eritrichium aretioides).* Perhaps the most striking example of miniaturization is seen in the Alpine Sunflower *(Hymenoxys grandiflora).* Its 2- or 3-inch flower heads are the biggest of all the tundra plants, yet it seldom reaches more than 4 inches in height.

Between here and the visitor center at Fall River Pass you will see other tundra plants as well. The road then descends to Milner Pass on the Continental Divide, where you should take time to see Poudre Lake. If it happens to be July there probably will be patches of snow on the meadows with carpets of Western Marsh Marigolds *(Caltha leptosepala)* and Globeflowers (*Trollius laxus* var. *albiflorus*) blooming by the thousands as it melts. A month or so later, you may be looking at such delights as Rosy Paintbrush *(Castilleja rhexifolia)* and Rocky Mountain Fringed Gentian *(Gentiana thermalis)* instead.

Fall River Road

Before the opening of Trail Ridge Road in 1932, the only way to drive over the mountains was on the old Fall River Pass Road, and this route is now open to the public, in a westerly direction only, starting from Horseshoe Park. It is unpaved, narrow, and steep and has hairpin turns and switchbacks, but it is safe and offers the visitor an enjoyable experience, especially since it can be combined with Trail Ridge Road to form a circular loop. An excellent explanatory booklet makes it an instructive self-guiding tour.

In its 9.4 miles, the Fall River Road climbs gradually for approximately 3,200 feet, going through the montane and subalpine zones, and eventually into alpine tundra. There are places where you can stop and relax in pleasant meadows, and seek out interesting plant species such as Star Gentian *(Swertia perennis)* or Rose Crown *(Sedum rhodanthum).* After you pass timberline you may see the related *Sedum roseum,* or Kings Crown, a dark red species. Among the many tundra flowers near the end of the drive is the large-flowered Arctic Gentian *(Gentiana romanzovii),* whose odd coloration very nearly approaches black and white.

GRAND TETON NATIONAL PARK

Composed of ancient rocks but torn from the earth by violent upheavals in the geologic yesterday, the Grand Tetons are a classic example of fault-block mountains. Their eastern face was the edge of a block that was thrust upward behind a fracture in the earth's crust about nine million years ago. Scoured clean of sedimentary layers by powerful glaciers, it now presents a steep granite facade of jagged pyramidal peaks reaching as high as 13,770 feet above sea level. If the Tetons seem more imposing than other western mountain ranges, this may be due to the fact that there are no foothills; they ascend directly from the floor of the valley for well over a mile. The drama of the scene is further enhanced when they are reflected in the placid lakes that lie at their base.

ARCTIC GENTIAN
Gentiana romanzovii

Grand Teton National Park is located in northwestern Wyoming, and is about the same size as Rocky Mountain National Park. The mountain range is in the western half, the major lakes are in the middle, and the Jackson Hole valley is in the eastern part. The principal road through the Park, U.S. 191 (also known as the Jackson Hole Highway), runs north from Jackson, paralleling the Snake River and connecting the visitor centers at Moose and at Colter Bay in the lower and upper sections of the park, respectively. For much of this distance the Teton Park Road provides an alternative route closer to the mountains. Access to the park from the east is by U.S. 26-287.

The southern approach to the Grand Tetons is through flat, arid country with log fences lining the roads and the gray pinnacled mountain range towering in the distance.

Much of the Jackson Hole country is dominated by Sagebrush (*Artemisia tridentata* along with several lesser species). Yellow is the prevailing flower color here, appearing in such plants as Rabbitbrush *(Chrysothamnus nauseosus),* Sulphur Flower *(Eriogonum umbellatum),* Gumweed *(Grindelia squarrosa),* and Prairie Coneflower *(Ratibida columnifera),* also called Mexican Hat. A few other hues are provided by the deep blue Low Larkspur *(Delphinium nelsoni),* the scarlet of Indian Paintbrush (Wyoming's state flower), and the variable Elk Thistle *(Cirsium foliosum).* Two alien species have invaded the roadsides: Nodding Thistle *(Carduus nutans),* with powderpuffs of bright purple florets, and Henbane *(Hyoscyamus niger),* whose petals are covered by an intricate network of dark veins. As pretty as any wildflower in their own way are the lustrous long-awned heads of Foxtail Barley *(Hordeum jubatum)* that grace the shoulders.

Excellent views of the mountains can be had from a number of points on both park roads—and especially from Jackson Lake Lodge—but for an eagle's-eye view of the extensive plains as well you should drive up Signal Mountain. There are scenic turnouts along the 5 miles to the summit, from which you look down on the valley from a height of a thousand feet.

The areas just north of here—Oxbow Bend, the tiny Christian Pond, and the larger Emma Matilda and Two Ocean lakes—are among the most beautiful in the park and comprise excellent wildlife habitat where moose, otter, and trumpeter swans may be seen. There are interconnecting foot trails through Lodgepole Pines and Aspen groves, and my notebooks record a galaxy of wildflowers in the vicinity, including Scarlet Gilia *(Ipomopsis aggregata),* Western Fringed Gentian *(Gentiana thermalis),* yellow Stonecrop *(Sedum stenopetalum),* the large-flowered Wild Flax *(Linum lewisii),* and Yampa *(Perideridia gairdneri),* plus many others.

The largest body of water in the park is Jackson Lake, but most of its trails are confined to the area south of Colter Bay; the 1.8-mile Colter Bay Nature Trail is an excellent lesson in the natural history of the region. Jenny Lake, Leigh Lake, and the connecting String Lake are at the foot of the mountains and are less crowded. Their shores are laced with footpaths where you are likely to find, among others, numerous ericaceous species such as Pink Pipsissewa *(Chimaphila umbellata),* Pinedrops *(Pterospora andromedea),* and the red-fruited Grouse Whortleberry *(Vaccinium scoparium).*

The boat dock on the west shore of Jenny Lake is the starting point for the Cascade Canyon Trail, which surely must be one of the most popular of all hikes (a motor launch will take you across). Many elect to hike on to Lake Solitude, which is 7.2 miles from the dock, but just a few hours' walk part way up the glacial valley will be amply rewarding.

On the far side of the lake you begin to climb gradually, and in a half-mile arrive at Hidden Falls, a 200-foot cascade that reveals itself only to those who hike the trail. Although the path ascends steadily, you are pleasantly distracted by the almost human, open-mouthed whistles of marmots doing sentry duty and the excited scurrying of diminutive pikas. It was here that I once was bending over the top of a tall plant, trying for a close-up photograph of the flowers and wondering why I was unable to keep it in focus, when suddenly the whole thing was whisked out from beneath the camera. Incredibly, a pika had been working on it inches away from my feet and finally severed it before I could trip the shutter. Scrambling over a talus slope, I eventually found the plant spread out with others on a flat rock where the pika was busily making hay to be stored underground as winter provisions.

Soon after Inspiration Point, with its panorama of Jenny Lake and the plains of Jackson Hole beyond, the terrain levels off and you pass through subalpine meadows of indescribable beauty. At one place there will be large stands of the handsome Pink Monkeyflower *(Mimulus lewisii)* with Indian Paintbrush *(Castilleja miniata),* Tall Chiming Bells *(Mertensia ciliata),* Balsamroot *(Balsamorhiza sagittata),* and Fleabane (*Erigeron* sp.). Cool shady spots will have intense ultramarine Mountain Gentians *(Gentiana calycosa);* more open ones, the delicate Colorado Columbine. Where meltwater trickles down from patches of snow, there is Fringed Grass-of-Parnassus *(Parnassia fimbriata).*

Woody plants are hardly less interesting. The black fruits of Twinberry Honeysuckle contrast with the red berries of Utah Honeysuckle *(Lonicera utahensis).* In the Rose family there are pink-flowered Subalpine Spirea *(Spiraea densiflora)* and Thimbleberry *(Rubus parviflorus),* close by a herbaceous St. John's-wort *(Hypericum formosum).*

This is, of course, only one part of what the eye is called upon to record. Rows of spirelike spruces and firs march up the rock-strewn, curving sides of this valley. The magnificent peaks that we have passed rise to awesome heights behind us, their flanks marked with snowfields and glittering waterfalls. Downslope, moose can be seen browsing or resting in the willow flats.

Eventually the trail enters a near-alpine environment (at 9,035 feet, Lake Solitude is not quite at timberline), and this will be evident from the vegetation, which includes such plants as Glacier Lilies *(Erythronium grandiflorum),*

American Bistort *(Polygonum bistortoides),* and Ballhead Sandwort *(Arenaria congesta).*

Another way to see the Grand Tetons is to take a 10-mile guided raft trip down the meandering Snake River within the park (not to be confused with hair-raising white-water adventures on rivers like the Colorado). These relaxing excursions afford an opportunity to view wildlife along the river, and to study the trees that grow on its banks, among which are Black and Narrow-leaf Cottonwoods (*Populus trichocarpa* and *P. angustifolia*), Blue Spruce *(Picea pungens),* and Whiplash Willow *(Salix caudata).*

YELLOWSTONE NATIONAL PARK

John Colter, who left his name on maps of the Grand Tetons, must have had a reputation for knowing what he was talking about, for Lewis and Clark took him on as a guide and trapper in 1806. Another mountain man, Jim Bridger, probably knew the West better than anyone in his day. But when these men told of finding strange country full of boiling ponds, mud volcanoes, mountains of glass, and fountains that spurted columns of hot water, they were confronted by disbelief and derision. Ultimately they were vindicated, after others went to see for themselves, became convinced, and returned to promote legislation to protect those very wonders of nature whose existence had been thought impossible. This culminated in the establishment of Yellowstone National Park in 1872.

In addition to being our (and, in fact, the world's) first national park, Yellowstone can claim many more superlatives. Its spectacular assemblage of thermal features—estimated at ten thousand—is unequalled anywhere. It is also our biggest park, covering 3,472 square miles. And Yellowstone Lake, which lies within its boundaries, is the largest mountain lake on the continent.

Yellowstone fills out the northwest corner of Wyoming just north of Grand Teton, edging into Montana and Idaho only slightly. There are five entrances: from the south via U.S. 191-89, from the west via U.S. 20 or 191 at West Yellowstone, from the north via Gardiner and U.S. 89, from the northeast via Cooke City and U.S. 212, and from the east via U.S. 14-16-20 from Cody. All of these routes lead to the 143-mile figure-eight Grand Loop Road within the park, which takes you to all of the major attractions including Old Faithful, Norris Geyser Basin, Obsidian Cliff, Mammoth Hot Springs, the Grand Canyon of the Yellowstone, Upper and Lower Falls, and Yellowstone Lake. You will probably see wildlife from the road also, especially in early morning or late afternoon; inquire as to the best places to see elk, moose, pronghorn antelope, bison, osprey, and trumpeter swans.

There are literally a thousand miles of hiking trails here, and although most of these go into the backcountry there are plenty of shorter ones as well, and botanizing can usually be made a part of visiting the hydrothermal points of interest. (It would be remiss to fail to acknowledge here that much of the beauty of Yellowstone's brightly colored hot springs is due to plants—different varieties of algae, each with a tolerance for a certain range of water temperatures and each imparting a distinctive hue.)

Take the Firehole Lake Drive, for example, to see masses of deep blue Western Fringed Gentians *(Gentiana thermalis),* the aptly named official flower of Yellowstone National Park, contrasting vividly with Woolly Yellow Daisies *(Eriophyllum lanatum).* Near Castle Geyser in the Old Faithful area, hundreds of Water Buttercups *(Ranunculus aquatilis)* hold their little white flowers just above the surface of the Firehole River. Farther north at Norris Geyser Basin there is a trail through Lodgepole Pines, with conspicuous growths of Dwarf Mistletoe (*Arceuthobium* sp.) on the branches and Ladies' Tresses *(Spiranthes romanzoffiana)* beneath, and the area around a cluster of small lakes known as Beaver Ponds near Obsidian Cliff is worth investigating for summer flowers.

Between Mammoth Hot Springs and the north entrance the elevation is considerably lower, as will be attested by such plants as Rocky Mountain Juniper, Rabbitbrush, and Prickly Pear Cactus *(Opuntia fragilis).* The Sagebrush flats in this section of the park are excellent for spotting pronghorns and coyotes.

Canyon Village is the focal point for viewing the canyon and falls, and there are walking trails along both rims with optional side trips down the steep sides. North of here many wildflowers can be seen during the summer months, among them Fringed Grass-of-Parnassus, Chiming Bells, Indian Paintbrush, Pink Monkeyflower, Willow Weed *(Epilobium latifolium),* and Engelmann Aster *(Aster engelmannii).* From Dunraven Pass there is a trail that climbs 1,380 feet in 5.5 miles to the top of Mount Washburn through pine forests and subalpine gardens. Near the 10,243-foot summit, which affords an overall view of the park, you may catch a glimpse of wary bighorn sheep.

GLACIER NATIONAL PARK

In 1932 the governments of the United States and Canada established the Waterton/Glacier International Peace Park, thereby subordinating the arbitrary political boundary represented by the 49th parallel to the natural unity of a magnificent million-acre wilderness. Each of the two countries administers its component: the 1,583-square-mile Glacier National Park in Montana and the smaller Waterton Lakes National Park in adjacent Alberta. The American section can be reached via U.S. 2 or U.S. 89.

The alpine scenery of Glacier National Park probably elicits the adjective "majestic" more than any other. Not surprisingly, glaciers are still a major feature, and although their numbers are decreasing many are accessible to hikers.

From June until mid-October, the very heart of the park can be visited by everyone, thanks to the 50-mile-long Going-to-the-Sun Road. A marvel of engineering and construction, this highway joins West Glacier at the lower end of Lake McDonald with St. Mary in the Blackfeet Indian Reservation on the eastern border. There is available a motorist's guide, which explains points of interest along this road and is of help in locating trailheads.

Since everyone who can drives at least part of the Going-to-the-Sun Road, a popular starting point for hikers is Logan Pass, where at 6,664 feet it crosses the Continental Divide. Going north from here, a 7.5-mile walk brings you to the Granite Park Chalets, where rustic overnight accommodations are available during the summer. Or you may choose to return the same day; it is an easy contour trail with no change in elevation. Along the entire route you can look across to the west at the imposing Garden Wall, a knife-edged arête left by a pair of parallel glaciers. Trailside wildflowers come in many shapes and colors, and are especially abundant around seeps where the ever-present Pink Monkeyflowers and Grass-of-Parnassus are joined by Mountain Death Camass *(Zygadenus elegans),* Globeflowers, and Buttercups. In drier situations are Silverleaf Phacelia *(Phacelia leucophylla),* Dryad *(Dryas octopetala),* several Penstemons, and the uncharacteristic *Heuchera cylindrica,* which goes by the name of Poker Alumroot. More widespread than most of these is Blanketflower *(Gaillardia aristata),* with coloring that suggests the inside of a peach. Unofficial trademarks of Glacier National Park, and impossible to overlook, are the tall heads of Bear Grass *(Xerophyllum tenax).*

For a much shorter walk from Logan Pass, you might try the 2-mile nature trail in the opposite direction to Hidden Lake Overlook. This begins in a sun-drenched alpine meadow replete with Yellow Monkeyflowers *(Mimulus guttatus),* Valerian *(Valeriana sitchensis),* Brook Saxifrage *(Saxifraga arguta),* and Elephant Heads *(Pedicularis groenlandica).* Rounding a turn at the end of the trail, you are suddenly treated to one of the most breathtakingly beautiful views in any western park.

Among other walks that can be taken from the Going-to-the-Sun Road is the Trail of the Cedars, a 1-mile nature trail that starts from the east side of the road north of the Avalanche Campground (park at the nearby picnic area). Here you can experience a fine example of a mature Western Red Cedar *(Thuja plicata)* forest with the white-flowered Queencup Beadlily *(Clintonia uniflora)* taking advantage of sunlit patches. Along Avalanche Creek, which is crossed by the trail, there are Western Hemlocks *(Tsuga heterophylla)* and Black Cottonwoods.

Other fine trails may be found almost anywhere, but another good location is the Many Glacier area. This is a 21-mile drive from St. Mary by way of the little settlement of Babb, Mont., a trip that would be worthwhile if it were taken only to see the picturesque old Many Glacier Hotel in its superb setting on Swiftcurrent Lake. From here a delightful walk of about 6 miles round trip passes by two smaller lakes and ends at Red Rock Falls. Western Serviceberry *(Amelanchier alnifolia)* is plentiful, and if you are here in August you should be able to feast on the sweet, juicy fruits, which are exceptionally large. Another abundant shrub, whose berries are edible but might better be left for the grizzlies, is Twinberry Honeysuckle; its specific epithet, *involucrata,* refers to the showy red bracts that open up to reveal the dark purple fruits. Among the more common herbs are Pink Pussytoes *(Antennaria rosea),* Globemallow (*Spheralcea* sp.), and Elk Thistle.

A somewhat longer hike from Many Glacier will take you to Iceberg Lake, an enchanting place where even in midsummer you can watch little icebergs calving and floating out into the turquoise water. Here there are cold climate plants like Mountain Heather *(Phyllodoce empetriformis),* Coolwort Foamflower *(Tiarella unifoliata),* Pink Pyrola *(Pyrola asarifolia),* and Siberian Chive (*Allium schoenoprasum* var. *sibiricum*).

Glacier National Park is home to a good many "big game" animal species. At Logan Pass, and elsewhere along the Going-to-the-Sun Road, be sure to scan the rocky cliffs with binoculars for shaggy white mountain goats, which are fairly common. Even though the total population of grizzly bears in the American Rockies numbers less than a thousand, there are enough in Glacier National Park to make an accidental encounter a distinct possibility. Although this knowledge should not interfere with your enjoyment of the park, it only makes sense to inquire beforehand as to any recent sightings along the trails you intend to use, and to follow whatever recommendations the rangers might make.

21

The Southwestern Deserts

IF THERE IS any single characteristic that is possessed by all deserts, it is aridity. We are too accustomed to the expression "dry as a desert" to be surprised by that statement until we are told that vast frozen regions of the North also are considered in a broad sense to be deserts because the moisture there is locked up in the form of ice and snow and therefore is *physiologically* unavailable to most organisms. By that definition, deserts are not limited to places like the Sahara, Arabia, or Australia, but are also to be found in less obvious situations on every one of the continents.

The deserts we are concerned with, however, are concentrated in the southwestern corner of the United States. There topography and storm patterns combine to severely limit the amount of rain (or snow), and warm latitudes, low elevations, and cloudless skies help the sun to dissipate the scant moisture that does fall. These factors have produced the dry, hot, sandy landscapes that to most of us represent the typical desert settings of western movies.

At first we are more aware of the earth itself than its cover of vegetation, but the paucity of plant life proves to be more perceived than actual. This is especially true of the Sonoran Desert, which covers much of southern Arizona and extends into California (where it is bisected by the Colorado River and because of this is often referred to as the Colorado Desert). The Sonoran also reaches far south along both sides of the Gulf of California.

Above the Colorado Desert, and confined mainly to southeastern California and the southern tip of Nevada, is the Mojave Desert. The flora of the Mojave is strikingly different from that of the Sonoran but equally fascinating.

The Great Basin Desert, which covers much of Nevada and Utah and projects into several adjacent states, is the largest in the United States but the least exciting botanically. (The Colorado Plateau, which has greater plant diversity, is sometimes considered a part of the Great Basin Desert but is actually a "semi-desert" and appears in Chapter 22.)

The fourth of the North American deserts lies far to the southeast and is isolated from the others. This is the Chihuahuan Desert, and like the Great Basin, it is primarily a shrub desert. Situated mainly in old Mexico, it protrudes slightly into New Mexico, but most of the U.S. portion is in the remote Big Bend area of Texas.

For plants to survive in a desert they must be well adjusted to cope with the austere environment, and no group exhibits such adaptations more dramatically than those belonging to the Cactaceae, or Cactus family. All Cacti are indigenous to the Americas, but some have been introduced into Europe, Asia, and Australia.

The all-important functions of storing and conserving water are perfected in the Cacti. Stems are greatly enlarged to accommodate masses of absorbent tissue, and are covered by an epidermis with ribs or knobs to allow alternate swelling and shrinking. Leaves, which would waste water through evaporation, are greatly reduced or entirely eliminated. Many species have clusters of spines, which not only protect the plant from predation by animals desperate for water but mitigate the drying effect of the hot sun by casting a diffuse pattern of shade on the stem surface. Cactus flowers are large and showy, with numerous intergrading petals and sepals, and most blossoms are short-lived.

It has been stated that of all our natural ecosystems the American deserts have suffered least at the hands of developers, and in general this may be true for they are somewhat short on allure and are anything but tractable. We can, however, find no consolation in the current practice of irresponsible commercial collecting of cacti on a massive scale, which has made them one of the world's most endangered major groups of plants. The demand for these and other desert succulents has reached the dimensions of a craze, not only in this country but in Western Europe, Japan, and even the Soviet Union. Protective statutes are few, and where they do exist are extremely difficult to enforce because of the immensity of the territory that needs to be policed.

Cacti and many other xerophytes are equipped with extensive networks of fine roots that enable them to take advantage of even the lightest shower and, as we have seen, give up this moisture with great reluctance. Other kinds of plants require abundant water but once this condition has been met can cope with the other rigors of desert life. Many must live near streams, even ephemeral ones, but all must send their roots deep into the earth and this has given them the name "phreatophyte," from the Greek word for "well."

Native plants that need to have their feet wet include Willows, Cottonwoods, Sycamore, Mesquite, and California Fan Palm, but none can approach the alien Tamarisk (*Tamarix* spp.) in the profligate consumption of water. Tamarisk is a deciduous shrub with bluish green scalelike leaves resembling those of juniper, and feathery narrow clusters of small pink flowers. It was brought here from its native Mediterranean region in the 1850s, and like all too many introductions, it quickly spread beyond control until it now covers more than a million acres. With an insatiable thirst it sinks its roots down to the water table, takes up enormous quantities of water, and transpires it just as fast, with the result that springs, pools, and streams are drying up to an alarming degree, to the detriment of indigenous flora and fauna. Another name for Tamarisk, Salt Cedar, alludes to the secretion through its leaves of salt crystals, which fall to the ground making the habitat even more inhospitable. Tamarisk has other properties that seem designed to thwart efforts toward eradication. It seeds itself prodigiously, grows rapidly, rebounds after fires, and resists herbicide sprays. Animals find its foliage unpalatable and cannot be bothered to eat the dustlike seeds.

The flavor imparted by mesquite to barbecued foods has become familiar all across the country, although many of its appreciators know little about the source of this fuel. The debt is owed to Mesquite, a spiny shrub common to all of our hot American deserts. It is a legume, with bipinnately compound leaves and clusters of yellow flowers that are followed by long, slender pods constricted at the intervals between the seeds. Several specific names have been given to these plants, but the genus is a confusing one and some taxonomists maintain that all mesquites are varieties of a single species, *Prosopis juliflora.*

Among the most dramatic displays of wildflowers anywhere in the world are the masses of so-called winter annuals, which burst into bloom in the hot American deserts between February and May. These spectacles occur only in some years, however. They are triggered by a complex combination of environmental factors, and although they cannot be reliably predicted, a series of gentle, well-spaced rains during the preceding winter is usually an encouraging sign. A second flowering of other species known as summer annuals may take place later in the year in response to the violent, heavy thunderstorms of July and August. The reason for these phenomena is quite simple: Lacking the mechanical means by which the more specialized desert plants manage to survive drought, these annuals resort to a completely passive form of adaptation by "sitting it out" until conditions are favorable for germination.

THE SONORAN DESERT

Perhaps the most impressive vegetation in the American deserts belongs to that part of the Sonoran known as the Arizona Uplands. This is where you find the big cacti, including the tall Saguaro, which the state has adopted as its official emblem.

Visitors to Phoenix can have their desert botanizing just about any way they like. For an easy and pleasant means of getting to know many of the larger plants, for example, you can take the Desert Foothills Drive. This is a horseshoe-shaped section of highway northeast of the city along which simple but conspicuous identifying signposts have been installed. From the junction of Pinnacle Peak Road with Cave Creek Road, drive north on the latter to the settlement of Carefree; then turn right on Tom Darlington Drive (Scottsdale Road) and continue until you again reach Pinnacle Peak Road. (Or go in the reverse direction if you prefer.) There are wide shoulders that enable you to safely pull off almost anywhere.

You will quickly become acquainted with such important members of this plant community as Giant Saguaro *(Cereus giganteus),* Palo Verde *(Cercidium microphyllum),* Mesquite *(Prosopis juliflora),* Desert Ironwood *(Olneya tesota),* Soaptree Yucca *(Yucca elata),* Jojoba *(Simmondsia chinensis),* Ocotillo *(Fouquieria splendens),* and Mormon Tea (*Ephedra* spp.). But you should also explore on foot a little distance back from the road for smaller, mostly unlabeled species. You may spot the sky blue flowers of Desert Hyacinth *(Brodiaea capitata)* atop long supple stalks, or the white Desert Chicory *(Rafinesquia neomexicana).* A leafless grayish shrub with tubular dull red blossoms being patronized energetically by hummingbirds is Chuparosa *(Beloperone californica).* And speaking of birds, keep an eye out for the handsome Gambel's quail, which are sure to be puttering about.

For a very different kind of drive, go up the Apache Trail (S.R. 88) from Apache Junction, which is on U.S. 60-89 about 40 miles east of downtown Phoenix. Only the first 20 miles of the road is paved, but even this segment will take you through beautifully awesome desert and canyon country beneath the towering Superstition Mountains. The road climbs and twists, and although there are frequent opportunities to get out and walk, they are limited to picnic areas and to turnouts provided specifically for parking.

DESERT CHICORY
Rafinesquia neomexicana

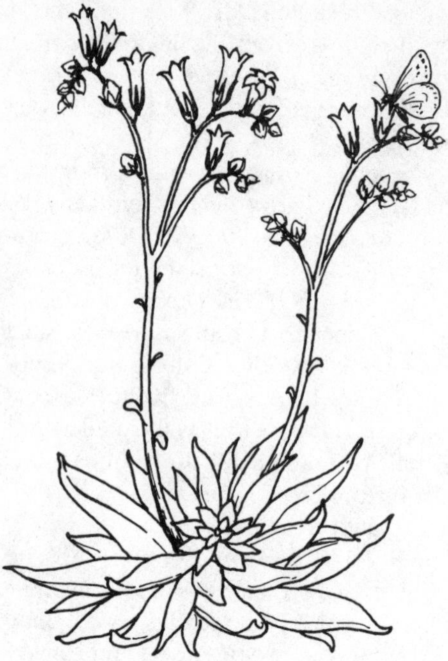

DUDLEYA
Dudleya sp.

Near the beginning there is an abundance of Palo Verde and Chainfruit Cholla *(Opuntia fulgida),* while farther up Brittlebush *(Encelia farinosa),* Prickly Pear *(Opuntia engelmannii),* and Mesquite become dominant and the roadsides are white with Fleabane *(Erigeron divergens).* Brittlebush is a very common desert composite, and is instantly recognizable by the way its yellow flower heads are held on branched stalks well above the dome-shaped mounds of silvery foliage. Growing out of tiny crevices in the rocky road cuts are succulents with yellow blossoms on red stems arising from rosettes of thick leaves—a species of *Dudleya.* On top of the banks are Hedgehog Cacti *(Echinocereus engelmannii)* and an occasional Century Plant *(Agave parryi).*

For those who want to combine hiking with their botanizing, Phoenix offers the world's largest municipal park, with an area of over 16,000 acres. South Mountain Park is a long ridge south of the city (at the foot of Central Avenue). Unfortunately, its 2,300-foot crest is marred by a cluster of more than 30 television antennae and associated structures, but it nevertheless contains superb desert foothills terrain.

Excellent paved roads make it possible to drive much of the park's length and enjoy spectacular views. There are extensive trails, and an especially attractive one that can be highly recommended for either a short or a long walk runs northeast from the parking area at Buena Vista Lookout. The trail passes almost immediately beneath some large boulders that are decorated with crustose lichens of a startling, almost fluorescent chartreuse intermixed with patches of mustard and rust. In the cool shade of these rocks there are colonies of Desert Tobacco *(Nicotiana trigonophylla),* a very viscid plant with white trumpet-shaped flowers.

The dominant tree is Foothills Palo Verde *(Cercidium microphyllum).* Its bark is an unusual moss green in color, and in early spring it becomes a mass of yellow flowers. Ocotillo *(Fouquieria splendens)* and Mormon Tea *(Ephedra trifurca)* with their odd growth forms provide visual accents. Ocotillo cannot possibly be mistaken for any other plant. A bundle of stems, long crooked canes sometimes forked near the top, ascend to form a vase-shaped shrub. They are covered with thorns, but for much of the time are leafless; following a period of rain they may put out small fleshy leaves, but with the next dry spell these will quickly wither and drop off. Tubular scarlet flowers appear in small clusters like bits of flame at the tips of the stems. A more prosaic plant commonly seen here is Desert Buckwheat *(Eriogonum fasciculatum),* a low shrub covered with small pink flowers.

Several kinds of cacti are in evidence, the spring blooming season beginning with the magenta-flowered Hedgehog. Among the many delicate herbs is *Lesquerella gordoni,* called Bladderpod because of the little inflated ball-like capsules that follow the butter yellow flowers and that pop when squeezed.

Driving south on I-10 toward Tucson, you will be sure to notice the distinctive shape of Picacho Peak, a remnant of the core of an ancient volcano, which rises almost straight up for 1,500 feet and is the site of Picacho Peak State Park. It can be climbed by a trail leading to the top, but most visitors elect to meander along the lower trails or drive the two short but pleasant scenic drives.

A self-guiding nature trail starting at the park office provides an excellent means of identifying and learning about many desert plants.

In early spring following favorable winter rains, an especially dazzling sight is provided by fields of Gold Poppies *(Eschscholtzia mexicana)* against the dark backdrop of Picacho's steep cliffs.

Saguaro National Monument

In the Tucson area the federal government has designated two tracts of desert scrubland as the Saguaro National Monument, preserving examples of the Giant Saguaro *(Cereus giganteus)* that to many is the very symbol of the Southwest. This imposing cactus, with its thick, ascending cylindrical arms reaching for the sky, may attain a height of 50 feet and an age of 200 years. Its seedlings require shade and shelter, and this explains why many a young Saguaro is seen growing very close to a Palo Verde or other "nurse tree." The Saguaro is exceptionally well adapted to collect and conserve water, which may constitute 90 percent of its mass. Vertical pleating permits it to expand when it rains and contract when it is dry, and rows of thorns break the force of desiccating air currents as do the

GIANT SAGUARO
Cereus giganteus

hairs of many other plants. The function of photosynthesis, which would be performed by conventional leaves if it had any, is taken over by the green columnar stems. Attractive many-petaled white flowers with a mass of golden stamens are borne at the extremities in summer.

The Rincon Mountain Unit ("Saguaro East") is located 17 miles east of Tucson, from which it is reached by way of Broadway Boulevard and Old Spanish Trail. Near the visitor center is the beginning of the 8-mile-long Cactus Forest Drive, a rather tortuous paved road with excellent views. Here too there are opportunities to walk away from the road, including the quarter-mile Desert Ecology Trail. The other section, known as the Tucson Mountain Unit or "Saguaro West," is 16 miles west of Tucson via Speedway Boulevard, Gates Pass Boulevard, and Kinney Road. A good 6-mile graded dirt road, the Bajada Loop Drive, traverses dense forests of Saguaro, and there are two nature trails as well.

Among the most common shrubs in the Saguaro National Monument are Triangle-leaf Bursage *(Ambrosia deltoidea)* and the inevitable Creosote Bush *(Larrea tridentata).* There are also a number of species of Cholla cacti, including Teddybear *(Opuntia bigelovii),* Buckhorn *(O. acanthocarpa),* Pencil *(O. arbuscula),* and Chainfruit Cholla *(O. fulgida).* The last is also known as Jumping Cholla because its joints separate from the parent at the slightest touch—

although the prudent hiker is less likely to brush against the plant than to walk into joints that have already fallen to the ground to take root.

Any visit to Saguaro West should include the Arizona-Sonoran Desert Museum, 2 miles nearer Tucson on Kinney Road. This widely acclaimed facility is a *living* museum in which all of the natural sciences are represented. Botanists will find displays of actual plant material typical of each life zone, a garden of cacti and succulents (including the incredible Boojum Tree, *Fouquieria columnaris,* from Baja California), demonstrations of landscaping with native desert plants, and colorful mixed beds of Owl Clover, Poppies, Verbenas, and Penstemons.

Organ Pipe Cactus National Monument

Organ Pipe Cactus National Monument occupies 516 square miles in southwestern Arizona, its southern edge coinciding with the Mexican border for 30 miles. Access is by S.R. 85, which runs south from I-10 west of Phoenix.

Within its boundaries is virtually the entire U.S. population of the giant cactus *Cereus thurberi,* which gets its common name from an imagined resemblance of its numerous cylindrical stems, which arise directly from the base, to the pipes of a large theatre organ. These green branches have many ribs armed with clusters of spreading thorns, and if you stand close on a windy day you can hear the breeze sing as it vibrates the multitudinous spines.

Two major auto loops are provided. The shorter is the 21-mile Ajo Mountain Drive, which winds among the foothills with a close look at the Ajo Range (which separates the national monument from the Papago Indian Reservation) and affords fine views of Organ Pipe Cactus. The Puerto Blanco Drive totals 51 miles, which includes an optional spur road to a population of the rare Senita Cactus *(Cereus schottii),* whose long gray whiskerlike bristles have given it the name Old Man Cactus, and a specimen of Elephant Tree *(Bursera microphylla).*

Both roads are excellent for seeing colorful spring wildflowers. Along the dry washes are the red *Penstemon parryi,* Purple Mat *(Nama hispidum),* Coulter's Lupine *(Lupinus sparsiflorus),* Fiddleneck *(Amsinckia tessellata),* and Chia *(Salvia columbariae),* with Owl Clover *(Orthocarpus purpurascens)* on grassy banks. Spread out on the surrounding plain are great numbers of Gold Poppy, Mojave Desert Star *(Monoptilon bellioides),* Desert Dandelion (*Malacothrix californica* var. *glabrata*), and others. Wild Heliotrope *(Phacelia distans)* and Pincushion Flowers *(Chaenactis steviodes)* share the shade of Creosote Bushes, while Desert Chicory seems to prefer growing within clumps of Bursage.

Many birds typical of the Southwest are common here, among them the Gila woodpecker, cactus wren, phainopepla, and roadrunner. Coyotes are frequently observed, and large bites out of Prickly Pear cacti attest to the presence of peccaries, or javelinas, which relish the "pads."

Algodones Dunes

Motorists traveling west from Yuma, Ariz., on I-8 pass through the brown sugar Algodones Dunes, where parallel roads facilitate flower hunting on either side. On the south the California government is developing access routes for off-road recreational vehicles, but the other side is less abused and its marginal road leads to such facilities as a plantation where the possible uses of the Jojoba tree *(Simmondsia chinensis)* are being investigated.

In good years these dunes are replete with such spring flowers as Desert Sand Verbena *(Abronia villosa),* Dune Primrose *(Oenothera deltoides),* Drooping Marigold *(Baileya pauciradiata),* and Stickleaf *(Mentzelia laevicaulis).*

Anza-Borrego Desert State Park

California has preserved a significant portion of the Sonoran Desert in its Anza-Borrego Desert State Park, the largest state park in the nation. The entrance, with an exceptionally fine visitor center, is located on Palm Canyon Road west of Borrego Springs.

Of the many hiking trails available, the one to Borrego Palm Canyon is probably the most interesting from a botanical viewpoint. It is 1.5 miles long each way, and is keyed to an interpretive pamphlet. The trail leads up an alluvial fan through a thick growth of Brittlebush, Trixis *(Trixis californica),* and Burrobush *(Hymenoclea salsola),* which is also called Cheesebush because of the odor given off by its twigs when the bark is scraped. Teddybear Cholla and Beavertail Cactus *(Opuntia basilaris)* are quite common, and herbaceous species include Desert Trumpet *(Eriogonum inflatum)* and the little Fremont Monkeyflower *(Mimulus fremontii).*

At intervals the path crosses Palm Canyon Creek, a refreshingly pretty stream in spring, its grassy banks studded with a scarlet Paintbrush (probably the wide-ranging *Castilleja linariaefolia*), but a bone-dry trough in summer. As the trail nears the canyon walls, new color is supplied by the emergent emerald foliage of Honey Mesquite and the red of abundant Chuparosa bushes. The climax is a native grove of California Fan Palms *(Washingtonia filifera)* among huge boulders, a cool, shady oasis that comes as a welcome change from the hot, mostly open trail.

Two other locations in Anza-Borrego are of botanical interest, especially for their numbers and variety of cacti. One is the higher slopes of the Yaqui Wells Nature Trail, off County Road S3 just north of the intersection with S.R. 78. The other is the Bill Kenyon Overlook Trail, 1.7 miles farther north on the same road. This is a one-half mile loop that features Desert Agave *(Agave deserti)* and affords a stupendous panoramic view of the San Felipe Wash and the surrounding mountains. Along these trails you can expect to find such attractive flowering

BEAVERTAIL CACTUS
Opuntia basilaris

species as Beavertail Cactus *(Opuntia basilaris),* Barrel Cactus *(Ferocactus acanthodes),* and Pincushion Cacti (*Mammilaria* spp.).

The most unusual plant in the park is the Elephant Tree *(Bursera microphylla),* a relative of Florida's Gumbo Limbo, chiefly known from Mexico. Its common name and its grotesque appearance are due to the fact that the trunks and branches taper much more rapidly than do those of other trees. Specimens have been planted near the visitor center, and others may be seen off Split Mountain Road.

The name of the park itself has an interesting derivation. Juan Bautista de Anza was a Spanish explorer who founded San Francisco in the same year that the American colonies declared their independence from England. "Borrego"—the Spanish word for lamb—refers to the peninsular bighorn sheep, a subspecies of which 400 individuals have found a refuge here.

Coachella Valley Preserve

To see the California Fan Palm at its best without strenuous hiking, you should visit the Coachella Valley Preserve, which is managed by the California Nature Conservancy. It is reached by exiting from I-10 (at Myoma, west of Indio) north on Washington Street; this curves west and becomes Ramon Road. At about 5 miles, turn right on Thousand Palms Canyon Road and drive for another 2 miles to the visitor center.

Like most other palm oases in the Southwest, this owes its existence to the San Andreas Fault, along which an upthrust barrier forces groundwater to the surface. A sandy 1-mile self-guiding nature trail parallels this bluff, and provides an excellent means of studying the characteristic shrub vegetation of the Colorado Desert.

The walk begins at Thousand Palms Grove in a dense concentration of the tall trees, which are the only species of palm native to the western United States. A distinctive feature is the skirt of dead fronds hanging down around the trunk, which may reach almost to the ground if not trimmed by fires. Always growing in association with them is Fremont Cottonwood *(Populus fremontii),* a water-demanding deciduous tree with furrowed bark and bright green foliage in the spring. To the Hopi Indians this was the traditional source of the wood from which they carved their kachina dolls. They would search through windrows of fallen trees washed up by flash floods and select those roots that had been aged for decades.

Plentiful along the trail is the purple-flowered Arrow Weed *(Pluchea sericea),* a shrub once used for arrow shafts. Other woody plants include Mesquite, Cheesebush, Four-winged Saltbush *(Atriplex canescens),* and Goldenbush (*Haplopappus* sp.). In several places the trail intersects wash beds, where Tamarisk and Smoke Tree *(Psorothamnus spinosus)* thrive. Even in dry weather there are some unpretentious but interesting small flowering plants in this sandy habitat, such as Sand Mat (*Euphorbia* sp.), Palafoxia *(Palafoxia arida),* and Desert Velvet *(Psathyrotes ramosissima).*

The trail ends in a loop enclosing a quiet pond, and winds through the McCallum Grove, a pristine stand of palms, before returning via the same route.

THE MOJAVE DESERT

North of the Colorado Division of the Sonoran Desert lies the Mojave (or Mohave), by far the smallest of the four American deserts if we include the Mexican portions as well. The Mojave is slightly higher and cooler than the Sonoran. This may seem paradoxical since it includes torrid Death Valley, which at one point is 282 feet below sea level, but we are speaking of averages, not absolutes. Both receive only scant precipitation, but whereas the Sonoran receives both gentle winter rains and brief summer storms, the Mojave is limited almost entirely to the former. This results in two characteristic patterns of vegetation: shrubs growing at widely spaced intervals and colorful displays of so-called winter annuals.

The most widespread plant in all our hot deserts is Creosote Bush *(Larrea tridentata),* which covers enormous areas often to the exclusion of any other species. In its ability to extract water from the soil it is almost too successful for its own good, for this leaves the surrounding area too depleted for even its own seedlings. It solves this, however, by sending out runners that, when they extend far enough, start new plants. Later the connecting stems, and eventually the original plant, will die leaving a circle of regularly spaced clones. This process is repeated over and over, with the result that an individual may have a lineage extending back through thousands of years. Aside from its prevalence, Creosote Bush is distinctive by its strong resinous odor and its olive green foliage set off by numerous yellow flowers.

The extravaganza of spring wildflowers that brightens the deserts following a season of favorable rainfall may be seen at its finest simply by driving through California's Antelope and Lucerne valleys, north of the San Gabriel and San Bernardino Mountains from I-5 east to Joshua Tree National Monument.

In a singular tribute to its official flower, the State of California has established a California Poppy Reserve in Antelope Valley approximately 15 miles west of Lancaster. The interpretive center is most easily reached by turning west from S.R. 14 on what is at first Avenue I, then becomes Lancaster Road (keep to the paved road). But despite the careful siting of this facility in what is considered the most consistent habitat of *Eschscholtzia californica,* it must be realized that there are years when even here there is not a single poppy flower to be seen. Such is the tenuous balance between winter rains and spring blooms.

Besides the California Poppy, certain other wildflowers are capable of creating homogenous blankets of a single color over whole acres in this region. One is Fiddleneck, whose yellow-orange flowers seem much too small until you consider the astronomical number of plants involved. Masses of bright purplish pink are due to an introduced member of the Geranium family known as Filaree *(Erodium cicutarium),* shown on the next page.

Many spectacular displays are made up of mixed composites, such as Tidy Tips *(Layia glandulosa),* whose yellow rays are bordered in white; Desert Dandelion, in which they are graduated from pale lemon to red at the center; and white Pincushion Flowers *(Chaenactis fremontii).* Other colors are supplied by purple Owl Clover, Notch-leaved Phacelia *(Phacelia crenulata),* and Pygmy Lupine (*Lupinus bicolor* var. *microphyllus*).

Joshua Tree National Monument

The 558,000-acre Joshua Tree National Monument is located north of California's Salton Sea, and except for the barren eastern portion is entirely within the Mojave Desert. It is in rugged and complex country, with imposing mountains and monoliths rising dramatically from the desert floor.

The Joshua Tree for which it is named is known scientifically as *Yucca brevifolia.* It is the largest of its genus, and its grotesque silhouette has come to symbolize the Mojave. The thick, awkwardly twisted branches end in tufts of

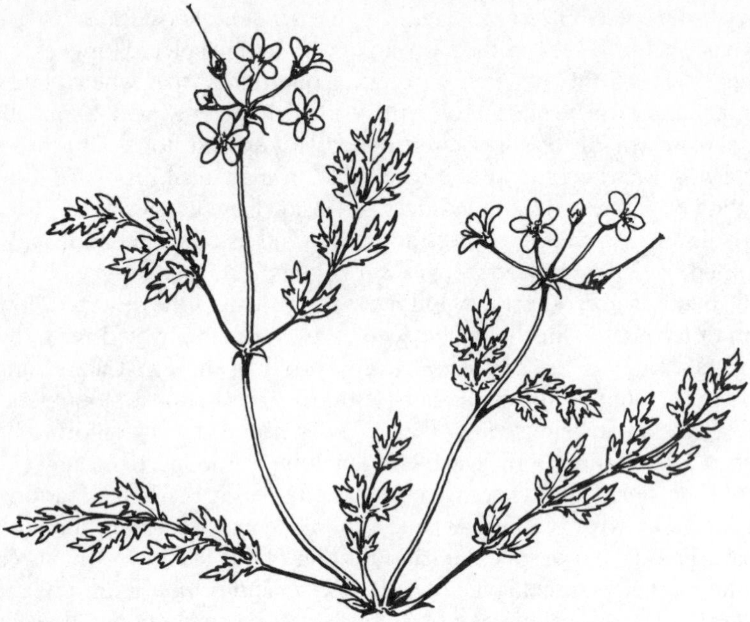

FILAREE
Erodium cicutarium

daggerlike leaves tipped with sharp spines, and in spring, but not necessarily every year, with large panicles of whitish flowers. Joshua Trees are by no means confined to this national monument (in fact, they occur only in the northern part) but are also plentiful in the valleys to the northwest.

A road runs through the preserve from the town of Twentynine Palms on S.R. 62 to a point on I-10 about 25 miles east of Indio, with other roads and hiking trails branching off in the northwestern sector and a visitor center at each end.

In the south, near the Cottonwood Springs Visitor Center, there is a small oasis with California Fan Palms, Fremont Cottonwoods, some introduced Willows *(Salix gooddingii),* and Mesquite trees bearing large clumps of red-berried parasitic Mistletoe *(Phoradendron californicum).* Below the spring is an arroyo where the water lies just a few inches beneath the surface of the sand.

As the road proceeds northward but is still at low elevations, the larger plants are Palo Verde and Chuparosa. Along the road margins there are large-flowered Canterbury Bells *(Phacelia campanularia)* and delicate Arizona Lupine *(Lupinus arizonicus),* and still farther on, the attractive yellow spikes of Prince's Plume *(Stanleya pinnata)* and Western Jimsonweed *(Datura meteloides).*

At one point a number of Smoke Trees share a dry wash with Pencil Chollas, and at another there is a large group of Ocotillos. About midway is Cholla Cactus

Garden, where the predominant species is Jumping Cholla, but there are many other desert plants on which a self-guiding trail provides information.

As the road gains in altitude and approaches the Joshua Tree community, *Isomeris arborea* becomes prominent. This shrub bears graceful yellow flowers with four petals flaring from a slender neck that produce green pods shaped like little punching bags, hence the common name of Bladderpod.

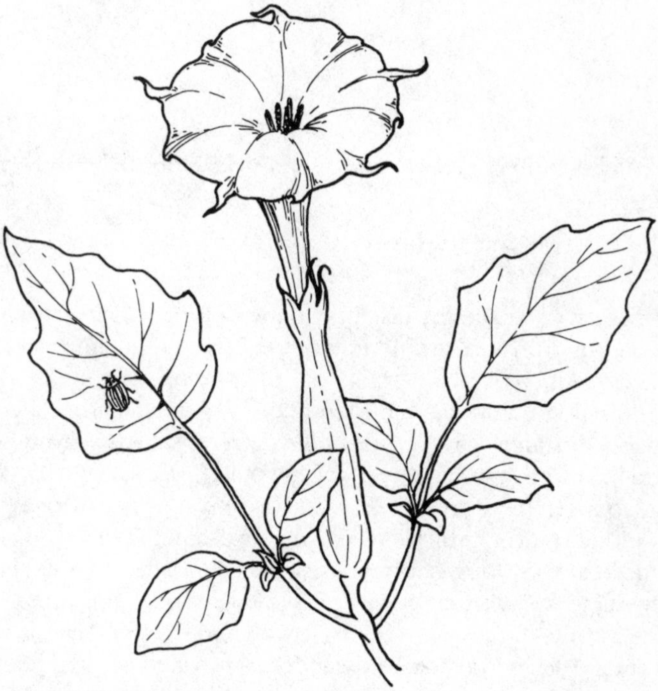

WESTERN JIMSONWEED
Datura meteloides

22

Canyon Country

AT DIFFERENT TIMES in the distant past the Colorado Plateau (named for the river, not the state) has taken the form of massive mountain ranges, flattened plains, ancient seabeds, and towering walls of layered sandstone, as the land was alternately uplifted, submerged, buried under sediment, and eroded by water, wind, and ice. Meandering rivers gouged irregular channels into the softer rocks and skirted the more resistant ones, which would become buttes and eventually wear away to resemble turrets, spires, and arches. A principal agent of erosion was the Colorado River, which created the awesome Grand Canyon and now flows a mile below its rim. Elsewhere, networks of smaller streams also followed lines of weakness, isolating blocks of stone and exposing them to further sculpturing by the forces of freezing temperatures, pelting rain, and sand-bearing winds. The spectacular landscapes that resulted are largely responsible for the fact that more than a dozen national parks and national monuments are located here.

The approximate location of the irregularly shaped Colorado Plateau can be indicated by tracing a 125-mile circle from a point slightly west of Four Corners, where Utah, Arizona, New Mexico, and Colorado meet. Because of relatively high altitudes the plateau is considered a "cold" desert (an assertion that summer visitors tend to challenge), and the flora is quite undesertlike (there are few cacti, for example). The tablelands are surrounded by rimrock, and behind this there are mountains of very considerable stature, and these capture the snows that account for most of the precipitation. Our attention, though, will be focused principally on the "uplands" between 4,000 and 7,000 feet. This corresponds roughly to the Upper Sonoran life zone, of which the indicator is a Juniper-Pinyon association.

The commonest shrub by far is Big Sagebrush *(Artemisia tridentata),* which surely must be the most famous plant in the West. This is the tallest of several species, sometimes reaching a height of 6 feet. The wedge-shaped, three-toothed leaves are covered with silky hairs, and these give it a peculiar gray-green color that contrasts vividly with the reddish soil. Ranchers loathe sagebrush because cattle are unable to digest it, and are especially frustrated by its ability to spread to new areas where grazing itself has removed competitive plants that otherwise would have held it in check. On the other hand, it is the pronghorn antelope's most important browse, and the sage grouse, for which it provides nesting sites, protective cover, and nearly all of its diet, probably could not survive without it.

CANYONLANDS NATIONAL PARK

My first reaction upon hearing of the establishment of Canyonlands National Park was to be turned off by the name. How, I asked, could the National Park Service, which always seemed to have been able to pick just the right appellation for each of its creations—Yosemite, Glacier, Big Bend, Acadia, Everglades—how could that agency possibly have coined such a Disneyesque name?

Any apprehension that Canyonlands might prove to resemble a theme park was, of course, immediately dispelled by reading its description. Here was a piece of truly rugged country, largely unexplored, and almost forbidding: the ideal place to spend a vacation without having to submit to the niceties so often associated with national parks. To make sure that our discomfort would be complete, it was decided that we should take our own car, which had neither air-conditioning nor four-wheel drive, that we would camp in a tent, and that (credit for this must go to our daughter's schedule at school) we would do this in the middle of the summer.

The Colorado River divides Canyonlands National Park in two, and our destination was the Squaw Flat Campground in the southern section, called The Needles. This is reached by a 38-mile entrance road that branches off to the west from U.S. 163 about 15 miles north of Monticello, Utah. Long before reaching the park boundary you come to an interesting roadside rock face covered with Indian petroglyphs, called "Newspaper Rock."

Incidentally, it is not at all necessary to camp out in Canyonlands. Large areas can be covered on perfectly good roads, and still more on hiking trails if desired. The places where roads become unsuitable for two-wheel-drive vehicles are clearly marked, and beyond this no one should venture without a guide unless fully experienced and capable of dealing with the possibilities of flash floods, quicksand, disorientation, scarcity of drinking water, and so on.

We decided to make camp near an overhanging shelf that was part of an immense red rock fortresslike formation, on the theory that during afternoon storms we could keep dry by sitting in the rear of our little nook and enjoy the view through a curtain of rainwater dripping from the ledge. What actually happened was quite different, for surface tension triumphed over gravity: The rain flowed inward over the sloping ceiling and waited to drip until the exact moment when it reached our backs.

The tent site itself was well shaded by a small grove of trees, which to no one's surprise turned out to be Pinyons and Junipers. The pinyon was the two-needled *Pinus edulis,* a name earned for it by its large, delicious seeds, which are harvested by Navajos and other tribes and prepared by a laborious procedure for their own use or for sale in Eastern markets as "pine nuts." Its companions were Utah Junipers *(Juniperus osteosperma).* These are upright but contorted trees branching from near the base. They are also called Desert Juniper, although their habitat is only semiarid, and locally they are known as "cedar," which of course is erroneous. They were adorned not only by an abundance of little fleshy cones, looking like bluish white berries, but by golden multibranched clusters of Juniper Mistletoe *(Phoradendron juniperum).*

The commonest flowering plant near the campsite was one of the Globemallows (*Spheralcea* sp.); they suggested miniature hollyhocks with cup-shaped flowers the color of tomato juice. A taller plant bearing large, dense clusters of yellow blossoms with long-protruding stamens was Yellow Bee Plant *(Cleome lutea),* a relative of the cultivated Spider Flower *(C. spinosa).*

Diffuse clumps of a plant with insignificant leaves and extremely delicate flowers proved to be *Ipomopsis longiflora.* The light blue star-shaped corollas with incredibly slender 2-inch-long tubes have given it the name of Pale Trumpets.

A day hike to Big Spring Canyon afforded an opportunity for distant views of many more of the fantastically eroded rock formations, and to watch an afternoon storm move along the horizon. The trail was no flower garden, but neither was it barren. Blossoming shrubs and herbs were frequent, and in the open environment it was impossible to overlook any of them.

One shrub that caught the eye because of its shape was *Gutierrezia sarothrae,* a many-branched composite that forms round cushionlike mounds studded with tiny yellow flowers. It has a high resinous content and burns furiously when ignited. Among its many common names are Turpentine Weed, Matchbrush, and Broom Snakeweed.

Several plants had obvious defenses against animal predation, one of which was *Coldenia palmeri,* a blue-flowered member of the Borage family. Its leaves, the edges of which were strongly rolled outward to reduce evaporation, were evenly covered with spiny bristles. Another no-nonsense plant was the dull red *Cirsium nidulum,* or Yellow-spined Thistle, one of several species with long, cylindrical flower heads.

Beyond any doubt the most handsome of the summer wildflowers was Desert Plume, or Prince's Plume *(Stanleya pinnata),* a plant several feet tall with gracefully curving spikes made up of myriads of small yellow flowers, their long stamens giving them a feathery appearance. But the prize for the most unusual must go to Desert Trumpet *(Eriogonum inflatum).* The *Eriogonums,* or False Buckwheats, number upwards of 150 species, most of which are in the South-

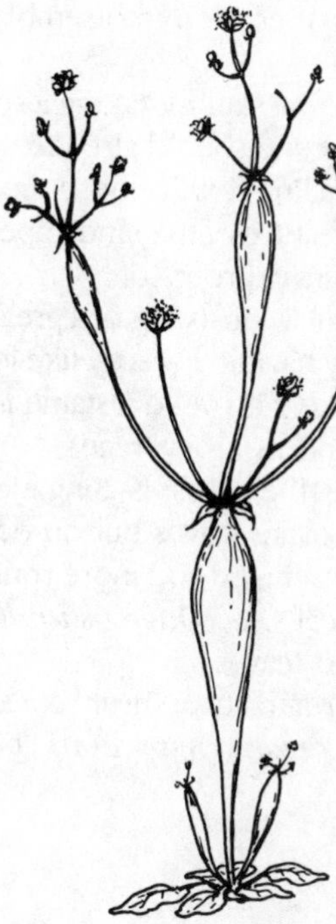

DESERT TRUMPET
Eriogonum inflatum

west and as a rule are rather unremarkable. In this example, however, the stems are greatly inflated just beneath the nodes, while the flowers are yellow and minute in small umbels above the bulbous swelling. This particular species was discovered by John C. Fremont, who was influential in freeing California from Mexican domination.

The northern portion of Canyonlands National Park, known as the Island in the Sky District, should be visited as well, if for no other reason than to experience the stunning aspects of the Colorado and Green River basins. A short spur off of the entrance road takes you to Dead Horse Point State Park, from which there is a breathtaking view of the tortuous, silt-laden Colorado far below.

ARCHES NATIONAL PARK

Just above the town of Moab, Utah, is Arches National Park, small in scale but superlative in its wealth of geological formations. It takes its name from the red sandstone arches, of which nearly a hundred have been discovered so far.

Arches are like natural bridges except that no stream is involved. They begin to form when cracks separate the rock into thin upright slabs called "fins"; further weathering then wears away the softer parts, creating windows at first, and eventually enlarging these into open arches. Other monoliths have been shaped by differential erosion into semblances of human figures, animals, and even city skylines.

There is much to be said for taking a self-guiding auto tour through the park on a paved road, but one should also take advantage of the foot trails in order to become better acquainted with the flora.

A conspicuous feature of the landscape is Mormon Tea (*Ephedra* sp.), which belongs to a primitive group of plants known as the Joint Firs. These are strange-looking shrubs consisting of a great many green branches at first appearing bare but actually having tiny scalelike leaves at the joints, or nodes. Staminate flowers are catkinlike with yellow stamens. Tea made from the stems has been used medicinally and as a beverage.

Unusual among the trees is Singleleaf Ash *(Fraxinus anomala),* which usually has simple ovate leaves but on occasion has two or three leaflets. The samaras are broadly winged and more rounded than those of other species. Also of interest is Wavyleaf Oak *(Quercus undulata),* in most locations a low shrub, with curly-margined leaves.

A common and attractive shrub consistently found in the pinyon-juniper zone is Cliff Rose *(Cowania mexicana).* Its flowers have five creamy white petals

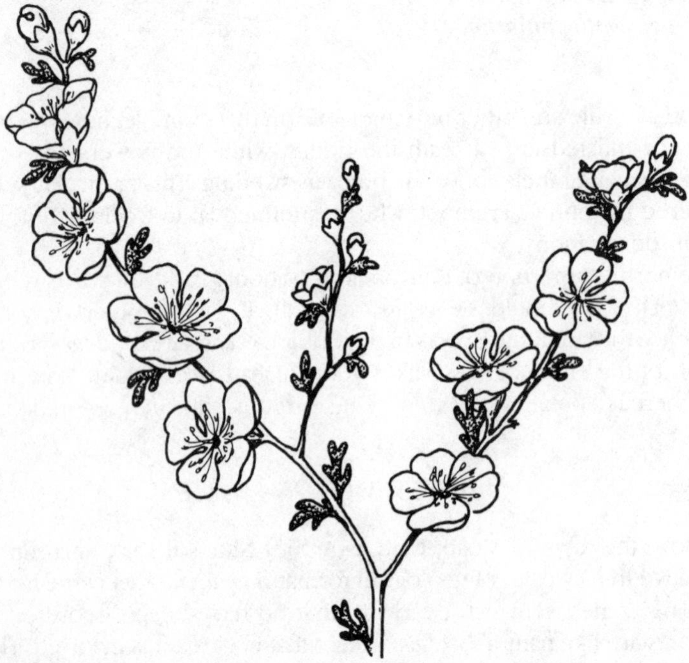

CLIFF ROSE
Cowania mexicana

and resemble small wild roses, but the leaves are small and deeply lobed. Their bitter taste accounts for another name, Quinine Bush, but does not deter deer, for which the foliage is an important food source.

If possible, no visitor should leave without taking the 1.5-mile trail from the Wolfe Ranch to Delicate Arch, an incredibly beautiful structure poised on the rim of a gigantic natural amphitheatre. To see it in the coppery glow of the late afternoon sun, framing the distant snow-covered peaks of the La Sal Mountains, is a sight never to be forgotten.

GRAND CANYON NATIONAL PARK

Of all the geologic wonders in the American Southwest, the one that put the word "canyon" in capital letters is, of course, the Grand Canyon of the Colorado. This extraordinary gorge of incredible beauty and magnitude is a full mile deep and has a maximum width of 18 miles between its jagged rims.

Only a relatively small segment of the Grand Canyon is included within the national park. Most visitors favor the South Rim, which is open all year (the North Rim, which is a 215-mile drive away, is higher in elevation and because of heavy snowfall is closed from late October until mid-May). An 80-mile drive on U.S. 180 from Flagstaff, Ariz., brings you to Grand Canyon Village inside the park boundaries, where there is a full range of services including the South Rim Visitor Center.

There are many ways of seeing the Grand Canyon. Hiking trails as well as mule trains will take you below the rim, and from bases outside the park you can raft down the river or fly over the canyon; each is rewarding in a different way according to your perspective.

To serve the less adventurous majority, there are two scenic drives with numerous parking overlooks along the South Rim. The West Rim Drive follows the edge of the canyon to Hermit's Rest, 8 miles from the village; to avoid congestion and excessive pollution this road is closed to private automobiles in the summer, but free shuttle bus service is provided as an alternative. The East Rim Drive goes in the opposite direction next to the Kaibab National Forest for 25 miles to Desert View. There are also several miles of foot trails along the canyon rim close to and paralleling these roads.

All motorists arriving at the South Rim get their first sight of the Grand Canyon at Mather Point, named in honor of Stephen T. Mather, the first director of the National Park Service. Here begins the struggle to comprehend the vast scale of the landscape—to imagine what forces could possibly have cut this 10-mile-wide gash out of stone, to believe that the tiny rivulet at the bottom is actually 300 feet across. One is instantly and totally absorbed in the scene, trying every viewpoint and peering in every direction, and even the most avid botanist will pay no heed to whatever flora may be near at hand.

It will not be much easier at the next lookout, Yavapai Point, although it might be difficult to ignore some of the shrubs if they happen to be in bloom. An attractive member of the Rose family, *Chamaebatiaria millefolium,* bears racemes of appealing white flowers and has finely dissected leaves that have earned it the name of Fernbush. Fremont Mahonia, actually a Barberry *(Berberis fremontii),* sports hollylike foliage and dense clusters of yellow blossoms. The glass-enclosed front of the museum at Yavapai Point offers a panoramic view down into the canyon, enabling one to pick out such details as mule trains on the path to Plateau Point and the Kaibab Suspension Bridge in the Inner Gorge.

After a stop at the visitor center in the middle of the village you may want to consider walking a part of the Rim Trail, which goes back as far as Mather Point and westward all the way to Hermit's Rest. The first part of the latter is recommended for early spring flowers, especially *Thlaspi fendleri,* popularly called Wild Candytuft, which sends its rounded tufts of white flowers up just a few inches above the ground under the Pinyon Pines. A companion plant, more spreading in its habit but no taller, is Dwarf Wood Betony *(Pedicularis centranthera)* with numerous beaked flowers splashed with bright red-purple at their tips. A very early blooming composite is Stemless Townsendia *(Townsendia exscapa),* an attractive large white daisy with a yellow disk that arises directly from a rosette of gray-hairy leaves. Still another low-growing plant is Wild Parsley (*Cymopterus* sp.), with crinkly bluish compound leaves and umbels of an unusual dark yellow hue.

Occasionally the tall thin spire of a Utah Agave *(Agave utahensis)* will be seen reaching as much as 20 feet into the sky, and contrasting sharply with the gnarled Utah Junipers clinging to the cliff edges. On a much smaller scale, Rockmat *(Petrophytum caespitosum)* fills rock crevices on and even below the

STEMLESS TOWNSENDIA
Townsendia exscapa

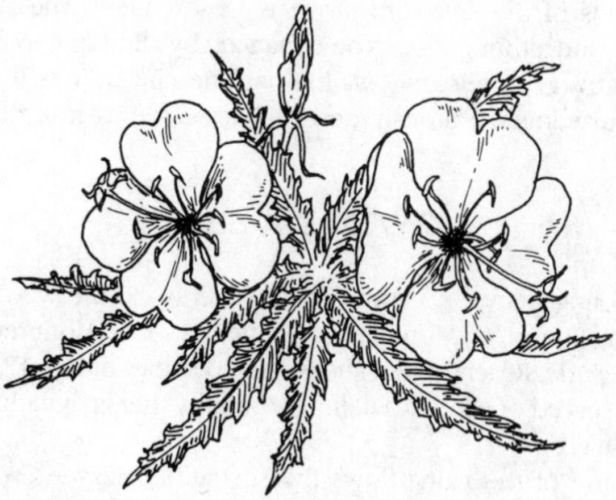

TUFTED EVENING PRIMROSE
Oenothera caespitosa

rim with masses of crowded foliage, from which dense heads of pink flowers will arise in late summer.

After a while the trail leads to a succession of hotels overlooking the canyon. In the vicinity of El Tovar—one of the most famous, built in 1904—look for purple drifts of a single species of wildflower with four narrow petals. This will be Perennial Rockcress, *Arabis perennans.*

You can, of course, continue walking along the trail, but if you wish to explore the area to the west you may prefer to drive to one or more of the overlooks. At Pima Point, for example, there is Desert Phlox *(Phlox austromontana)* in spring and white Tufted Evening Primrose *(Oenothera caespitosa)* somewhat later in the year.

Stands of Ponderosa Pine occur in some of the more protected locations in and around the village, and Gambel's Oaks are scattered here and there on the rim, but to see these trees in their typical habitat you might stop at one of the picnic areas spotted along the East Rim Drive. The stately *Pinus ponderosa* is an important source of timber. Older specimens have a distinctive orange bark that flakes off in thin, irregularly shaped scales and has a faint aroma suggestive of vanilla. *Quercus gambelii,* which has rather small leaves with deep rounded lobes, often forms dense thickets in these pine woods.

Each of the observation points on the East Rim Drive is likely to yield something different. At Grandview Point, Banana Yucca *(Yucca baccata)* is plentiful, while the gravelly, sparsely shaded area behind the parking space at Lipan Point is a good place for *Astragalus* (referred to locally as Locoweed) and Indian Paintbrush *(Castilleja chromosa).*

The terminus of the East Rim Drive is Desert View. The "view" is of the Painted Desert, and although it can be enhanced by climbing up inside the stone pseudo-Indian tower, a leisurely walk along the cliff trail as it winds through Sagebrush, Four-winged Saltbush, and Cliff Rose seems more appropriate.

MONTEZUMA CASTLE

For those traveling between Flagstaff and Phoenix on the way to or from the Grand Canyon, a stop at Montezuma Castle National Monument provides a refreshing interlude. Reached by a short spur road east of U.S. 17, it features one of the best preserved of all cliff dwellings, built by the Sinagua Indians between 1100 and 1400 A.D.

Near the base of the ruins flows the spring-fed Beaver Creek—at various times a purling stream, a hesitant trickle, or a rampaging torrent. Its floodplain is populated by fine white-limbed Arizona Sycamores *(Platanus wrightii),* the leaves casting a pattern of dappled shadows that repeat the patchwork bark above. Stone-hewn sycamore logs were used for ceiling supports in the cliff houses; the ends of these timbers can be seen protruding from the mud walls.

Winding through the bottomland beneath the ruins is the Sycamore Trail, along which many woody plants are identified by signs. Among them are Desertwillow *(Chilopsis linearis)* and Seepwillow *(Baccharis glutinosa),* both of which were used as roof-building material, as well as others that furnished food, medicines, and dyes.

23

The Sunset Coast

CALIFORNIA'S extravagantly varied thousand-mile Pacific coastline has something for just about everyone, and not the least favored are the natural historians. Its attractions for the botanist are many—as will be amply demonstrated as we travel along the shore with occasional digressions inland—but one of the most striking features is the large number of endemic or near-endemic species of evergreen trees.

Torrey Pine *(Pinus torreyana)* has the narrowest distribution of any pine in the United States. It has been given sanctuary in the Torrey Pine State Reserve north of La Jolla, and this is fortunate for us as well since its only other native habitat is on Santa Rosa Island, which is privately owned.

Grotesquely misshapen by strong winds and salt spray until they look like giant bonsai specimens, the much-photographed Monterey Cypresses *(Cupressus macrocarpa)* are the most widely recognized of California's trees. This species is the largest of the genus and is one of the world's rarest trees, occurring only in two small stands south of Monterey Bay. (Actually there are four other cypresses here, all uncommon and with limited ranges in California's coastal region, but they are smaller in stature and much less well known.) The foggy atmosphere that enhances the picturesque silhouettes of these trees as they cling to the rocky seaside cliffs may have contributed to their adaptation to this particular locale, by moderating both winter and summer temperatures and by providing moisture that, condensing on contact with the foliage, supplements the meager rainfall.

Growing on the hills behind the Monterey Cypress groves is Monterey Pine *(Pinus radiata),* another tree of restricted distribution. It has a spreading crown and persistent, obliquely ovoid cones bearing minute prickles. Bishop Pine *(Pinus muricata)* also has lopsided cones, but they are armed with stout spines.

This tree occurs in a number of widely separated localities on the coast both above and below San Francisco, as well as in Baja California.

One of the most unusual of the state's gymnosperms is California Torreya *(Torreya californica).* A member of the Yew family, it bears large-seeded plum-like fruits and is sometimes called California Nutmeg. Although not at all common, it is scattered widely along the northern coast and as far south as the Santa Cruz Mountains, as well as on the western slopes of the Sierra Nevada.

The most famous of all the coastal trees is, of course, the Redwood *(Sequoia sempervirens).* Its reputation for producing the world's tallest trees (the record holder measures 367.8 feet) has sometimes led to its being confused with the Giant Sequoia, or Big Tree *(S. gigantea),* whose champion specimens are more massive and have attained much greater age. Both belong to the same genus (they are the sole survivors) and both have red heartwood, but their habitats are very different. *S. sempervirens* grows only in a fog-shrouded belt near the coast; it has been heavily logged, and it is estimated that in a little more than a century 90 percent of the virgin trees have been removed. On the other hand, *S. gigantea* is restricted to a small strip on the western side of California's Sierra Nevada, where it is being resolutely protected.

All of the foregoing trees grow at low elevations, but other endemic species may be found on the coastal mountain ranges. One that occurs in the Santa Lucia Mountains near Monterey is Bristlecone Fir *(Abies bracteata).* At a distance this tree can be recognized by its thin, pointed spire. Other features by which it differs from the other western firs can be seen at close range: needles drawn out to a long sharp point, and fat oval cones with slender protruding bracts that spread conspicuously far beyond the loose scales.

Farther south, in the Santa Ynez Mountains behind Santa Barbara, is found Bigcone Douglas Fir *(Pseudotsuga macrocarpa).* A much smaller tree than the well-known Douglas Fir *(P. menziesii),* it has little economic value.

An interesting species of Pinyon extends from Baja California to the desert slopes of the Santa Rosa Mountains near Los Angeles. It is known as Parry Pinyon, and because its needles are usually in clusters of four, it was given the scientific name of *Pinus quadrifolia.*

The southern coastal mountains are home to the Coulter Pine *(Pinus coulteri),* which boasts the heaviest of all cones, armed with long, incurved claws. The stateliest of all pines, though, belongs more to the cool mountains than to the seacoast. It is the Sugar Pine *(P. lambertiana),* our tallest species and bearing the largest cones, which measure a foot and often more in length.

TORREY PINES STATE RESERVE

Much more than just a refuge for its namesake, Torrey Pines State Reserve is a veritable showcase for wildflowers of the southern California seacoast. It is easily reached from County Road S3 a mile south of Del Mar. A number of trails have been laid out, and maps and descriptions are available at the visitor center, which is housed in a picturesque adobe building.

For a starter, the Guy Fleming Trail is recommended. This is a 0.7-mile loop on level sandy paths, and includes a sizable stretch along high oceanside cliffs. Setting out in a counterclockwise direction, you are introduced to such diverse woody plants as a shrubby sumac *(Rhus integrifolia)* with the name of Lemonadeberry, California Scrub Oak *(Quercus dumosa),* and Mojave Yucca *(Yucca schidigera).* A small grassy seep has Shooting Star *(Dodecatheon clevelandii),* a large-flowered Blue-eyed Grass *(Sisyrinchium bellum),* and Indian Paintbrush.

At the North Overlook the Torrey Pines begin, their struggle with the elements made visible by the way they lean strongly to leeward. Although they appear tortured, they are well equipped to cope with the dry, salt-laden winds off the ocean, and they are far from puny: Their needles, which come in bundles of five, can approach a foot in length, and the robust cones half of that.

On both sides of the cliff trail, which looks down on the blue Pacific, are drifts of wildflowers growing low to escape being torn to shreds by the constant breeze. There are charmers like Suncups *(Camissonia bistorta),* Beach Sand Verbena *(Abronia umbellata),* Tidy Tips *(Layia platyglossa),* and Fringed Pink *(Gilia dianthoides).* Hottentot Figs *(Mesembryanthemum edulis),* gross succulents with thick three-angled leaves and showy yellowish flowers, and the similar red-purple Sea Figs *(M. chilense)* are here in solid masses but have not taken over such great expanses as, for example, at Carlsbad just a few miles to the north. More conspicuous because of their height are Coast Prickly Pears *(Opuntia littoralis).*

Turning back again from the sea, the path goes through taller growth, which includes Blue Nightshade *(Solanum parrishii)* and Bush Monkeyflower *(Mimulus puniceus).*

FRINGED PINK
Gilia dianthoides

SANTA ROSA PLATEAU PRESERVE

California's freshwater vernal pools are one of its ecological treasures—and one of the most threatened because they occur in fertile areas attractive to agricultural interests. These shallow depressions, which vary greatly in size, are lined with clay hardpan and hold winter rain longer than does the surrounding grassland. The advent of warm weather gradually dries up these pools while setting in motion the germination of millions of wildflower seeds. These flowers burst into bloom around the pond edges at first, then closer to the center as the water recedes, sometimes forming concentric rings of several different species.

Fortunately, The Nature Conservancy has made it possible to visit a series of vernal pools atop Mesa de Colorado in its Santa Rosa Plateau Preserve, which is especially convenient for travelers between San Diego and Los Angeles. To reach it, turn west from I-15 just north of Murrieta, on Clinton Keith Road. At 5 miles you will come to the Englemann Oak Woodland Trail (another feature of the same preserve, which you may wish to investigate), dedicated to preserving one of the last remaining viable stands of *Quercus engelmannii,* also known as Mesa Oak. From this point, continue for another 2.6 miles, first on Tenaja Road, then on Via Volcano, to the Vernal Pool and Overlook Trail. (Both are "walk-in" trails with interpretive pamphlets available at trailhead.)

The species of flowers forming dense carpets of color may vary, but some that can be counted on are Tidy Tips, Goldfields *(Lasthenia chrysostoma),* California Poppy *(Eschscholtzia californica),* Downingia *(Downingia bella),* Fiddleneck (*Amsinckia* sp.), Purple Owl Clover *(Orthocarpus purpurascens),* and

CALIFORNIA POPPY
Eschscholtzia californica

Fringed Pink *(Gilia dianthoides).* There are also patches of Popcorn Flower *(Plagiobothrys nothofulvus),* Miner's Lettuce *(Montia perfoliata),* a beautiful violet Shooting Star (*Dodecatheon* sp.), and many more. Along the road, you will notice the handsome Yellow Lupine, *Lupinus macrophyllus,* growing among the oaks.

POINT LOBOS STATE RESERVE

Point Lobos State Reserve is another California park whose concentration of a single rare conifer—in this case Monterey Cypress—would alone be sufficient to distinguish it but that offers a plenitude of interesting flowering herbs as a dividend. And as if this were not enough, it is a favorite place from which to observe sea lions, seals, sea otters, and migrating gray whales.

Occupying 1,325 acres on a rugged peninsula 4 miles south of Carmel, its entrance is on the west side of S.R. 1. The park is crisscrossed by a network of trails, none of them far from the park roads. Predictably, one of these is labeled the Monterey Cypress Grove Trail, and is the best place of all to see these gnarled, wind-twisted trees holding precariously to their yellow cliffs above the pounding surf. Others back from the edge have circular beds of slender Mountain Iris *(Iris douglasiana)* and masses of purple California Hedge Nettle *(Stachys bullata)* growing beneath them.

In open grassy areas there are Star Lily *(Zygadenus fremontii),* Checker Bloom *(Sidalcea malvaeflora),* tiny Coast Lotus *(Lotus formosissimus),* and *Sanicula arctopoides,* a strangely flattened plant called, rather appropriately, Footsteps of Spring. Some paths, such as the entrance to the Cypress Grove loop, are flanked by high growth that includes Wild Rose *(Rosa californica),* Beach Sagewort *(Artemisia pycnocephala),* and Poison Oak *(Rhus diversiloba),* but have as their showiest component the handsome blue-flowered Wild Lilac *(Ceanothus thyrsiflorus).*

In the interior of the peninsula, trees with dark, deeply fissured bark will be recognized as Monterey Pines *(Pinus radiata).* This is the same species as the famous "butterfly trees" at Pacific Grove (on Ridge Road off Lighthouse Avenue) to which multitudes of monarchs from the intermountain region flock on their winter migrations. Their reason for choosing this species exclusively is as much a mystery as why even greater numbers, representing the breeding population of the eastern two-thirds of the country, seek out a grove of evergreen trees known as *oyamel* in Mexico's Sierra Madre.

MONTEREY BAY AQUARIUM

For most of us, the only marine algae we get to observe comes in the form of fragments that have been torn loose and cast up on a beach, but this has now been remedied to some extent by the Monterey Bay Aquarium, a superb new facility located on Cannery Row in Monterey, Calif.

Its centerpiece is a dramatic three-story-high tank in which a forest of Giant Kelp *(Macrocystis pyrifera)* has been established. These plants and other seaweeds, including some with curious names like Sea Grapes *(Botryocladia pseudodichotoma)* and Turkish Towels *(Gigartina corymbifera),* are anchored by their holdfasts just as they would be in their natural environment and, together with the scores of fish that live among them, are maintained by a steady flow of seawater.

There also are tidepools to demonstrate the different intertidal zones, each with characteristic flora and fauna.

Another fascinating feature is an indoor aviary in which you can observe avocets, plovers, stilts, egrets, and other shorebirds literally at arm's length without any intervening glass or screen. Miniature sand dunes with some of their typical vegetation have been moved into this exhibit, which is complete even to the periodic lapping of waves upon the little beach.

MUIR WOODS NATIONAL MONUMENT

All of the world's Coast Redwoods *(Sequoia sempervirens)* grow in a 500-mile strip along the Pacific Ocean between California's Monterey County and the extreme southwest corner of Oregon. To say that the public has easy access to these tall trees would be an understatement, for there are 29 California state parks alone in the region, ranging all the way from 15 to almost 44,000 acres in size.

From all of the available choices, it would be hard to imagine a better example than Muir Woods for one's introduction to the redwoods and their environment. Only 17 miles north of San Francisco, it is reached by a loop road off S.R. 1. Although close to the city, Muir Woods actually owes its survival to being located in an isolated canyon from which it was not feasible to remove logs. At least that was the case in the 19th century, and by 1908—before the lumber barons could figure out a solution to the problem—President Theodore Roosevelt had proclaimed it a national monument.

Its 484 acres, together with the adjacent and much larger Mt. Tamalpais State Park, provide opportunities for extensive hiking, but the largest trees (250 feet high) are just a short walk from the visitor center. In fact, much of the typical redwood forest flora can be seen along the easy 1-mile Cathedral Grove Loop Trail.

It is immediately apparent—probably because of the stark contrast—that there are a number of other trees here, among them California Buckeye *(Aesculus californica)* and Bigleaf Maple *(Acer macrophyllum).* Tanoak *(Lithocarpus densiflorus)* is the only American representative of this Asiatic genus; its staminate flowers are catkins like those of chestnuts but shorter, and its fruits are

REDWOOD SORREL
Oxalis oregana

acorns. Instead of a balsamlike odor, which we might reasonably expect when surrounded by such enormous needle-bearing evergreens, there is a pervasive spicy aroma. This is due to *Umbellularia californica,* a small tree variously called Bay, California Laurel, and Oregon Myrtle.

Several species of wildflowers form solid masses where sunlight manages to reach the forest floor, notably Miner's Lettuce, Pacific Houndstongue *(Cynoglossum grande),* and the delightful pink Redwood Sorrel *(Oxalis oregana).* Others occur in shadier places and more sparingly: Milkmaids *(Dentaria californica),* Pacific Trillium *(Trillium ovatum),* Red Clintonia *(Clintonia andrewsiana),* the yellow Redwood Violet *(Viola sempervirens),* and a species of Wild Ginger with elongated calyx lobes, *Asarum caudatum.* Interesting for both its flowers and its name is *Tellima grandiflora,* or Fringecup, *Tellima* being an anagram of *Mitella,* the name of a similar genus.

As might be expected, the canyon is rich in ferns. The most conspicuous is Western Sword Fern *(Polystichum munitum)* on the steep banks above Redwood Creek.

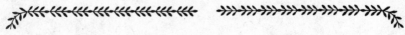

24

The Pacific Northwest

THE TOPOGRAPHY and climate of the Pacific Northwest are so varied as to defy characterization. Paralleling the coastline is the rugged Cascade Range, which has among its features the majestic, much-visited Mount Rainier, and the wild, remote North Cascades National Park adjoining the Canadian border. On the lowlands east of this range, where precipitation is greatly reduced, there are sagebrush plains and vast fields of wheat. Other mountains—the Olympics—crown the peninsula that bears their name, forming a barrier against the moisture-laden winds to create verdant rain forests on one side and a near-desert on the other. Even the waters display extremes: placid tidepools along the Strait of Juan de Fuca and thundering surf battering the seacoast. The wide variety of resources that contribute to the economy—among them timber, grain, orchard fruits and seafood—testifies to the natural diversity of the region.

MOUNT RAINIER NATIONAL PARK

Only rarely do we name a national park for a single mountain peak, but if there is any eminence that deserves to share that honor with Alaska's Mount McKinley, it is Washington's Mount Rainier. Rising to 14,410 feet above sea level, Mount Rainier comes within a mere 27 yards of being the tallest mountain in the lower forty-eight, but perhaps what is more impressive is the fact that for more than half of its height it towers majestically over the surrounding plateau, aloof from its neighbors and bedecked with no fewer than 26 named glaciers. Mount Rainier's conical form reveals that it once was an active volcano, but its last major eruption occurred thousands of years ago and today the mountain is quiescent and its glaciers stable.

To see this perpetually snow-covered summit looming against a cerulean sky is to experience one of the most beautiful spectacles in the West, but it is a sight sometimes denied to visitors when a cloud embraces it and is reluctant to leave. During one August that included a week's stay at Paradise Inn in Mount Rainier National Park, my only glimpses of the top were when arriving (from a turnout on the winding road partway up) and after leaving (from a plane window en route from Seattle to San Francisco). There were ample compensations, however, in the way the morning mists lightened the somber forests and condensed on the myriad wildflowers like glistening beads of glass.

The lowest slopes of Mount Rainier are home to the largest of its trees—species that we will see at their best in the rain forests of the Olympic Peninsula. Up to about 5,000 feet we pick up Western White Pine *(Pinus monticola)* and several true Firs: Grand, Noble, and Pacific Silver *(Abies grandis, A. procera,* and *A. amabilis).* An unusual and attractive tree is Alaska Cedar, or Yellow Cypress *(Chamaecyparis nootkatensis),* which has yellowish green pendulous branchlets and little round cones with pointed scales. Some of these trees persist into the Hudsonian zone, eventually being reduced to a stunted, sprawling form, which is also true of typical high-altitude species such as Whitebark Pine *(Pinus albicaulis)* and Subalpine Fir *(Abies lasiocarpa).*

It is in the narrow belt between 5,000 and 6,500 feet that we find Rainier's renowned subalpine meadows, which John Muir likened to a wreath of flowers encircling the mountain when he climbed to the summit a century ago. His metaphor was appropriate, and these natural gardens occur in a number of places on the slopes, but there can be no doubt that in terms of popularity the Paradise area is the clear winner. Situated at an elevation of 5,400 feet, Paradise Inn is flanked by groves of conifers interspersed with sunlit fields replete with flowers. In addition to this famous three-story building, ruggedly built to withstand snows that in some winters bury it to the roof, there is a network of trails including some that lead to ice caves and to superb views of glaciers.

In all, 700 species of flowering plants have been recorded in the park, but a much smaller number, blooming in great profusion and in successive waves, account for the great displays. In the Hudsonian zone these peak during July and August, when the Western Pasqueflowers *(Anemone occidentalis)* have only their tousled fruiting heads to show that they had blossomed earlier in the year. A few other plants are tall enough to stand above the rest, notably the pleated leaves of False Hellebore *(Veratrum eschscholtzii),* the dusty rose Subalpine Spirea *(Spiraea densiflora),* the stunningly beautiful blue and white Subalpine Lupine *(Lupinus subalpinus),* and Broadleaf Arnica *(Arnica latifolia).* Others compensate for their modest size with colors so rich as to suggest a stained glass window: Magenta Paintbrush *(Castilleja oreopola),* Fanleaf Cinquefoil *(Potentilla flabellifolia),* Skunkleaf Polemonium *(Polemonium pulcherrimum),* and Jeffrey Shooting Star *(Dodecatheon jeffreyi),* their brilliance relieved at intervals by the white puffs of American Bistort *(Polygonum bistortoides).*

Two low shrubs, both belonging to the Heath family, are often found growing in close proximity: red Mountain Heather *(Phyllodoce empetriformis)* and the white-flowered Mertens Cassiope *(Cassiope mertensiana).* As we progress upward toward the Arctic-Alpine zone smaller and more widely spaced plants are encountered, such as the red-stemmed Yellowdot Saxifrage *(Saxifraga austromontana)* and the dainty Partridge Foot *(Luetkea pectinata).*

Certainly it is understandable that the spectacular subalpine meadows should claim our attention, but it would be a mistake to overlook the flora of the lower elevations where the dense tree growth limits the numbers of flowering plants. These can easily be observed along the approaches to the Paradise area from either side.

One of the most conspicuous—and possibly the most distinctive—of western mountain wildflowers is Beargrass *(Xerophyllum tenax).* Nearly as high as a man, it carries a clublike raceme of small white flowers above a clump of extremely narrow leaves. Edible fruits are borne by Salmonberry *(Rubus spectabilis),* a shrub with showy pink blossoms, and by the ericaceous Salal *(Gaultheria shallon),* which has rows of urn-shaped flowers typical of the Heath family.

Among the smaller wildflowers are several orchids, including the Slender Bog Orchid *(Platanthera stricta),* as well as Coolwort Foamflower *(Tiarella*

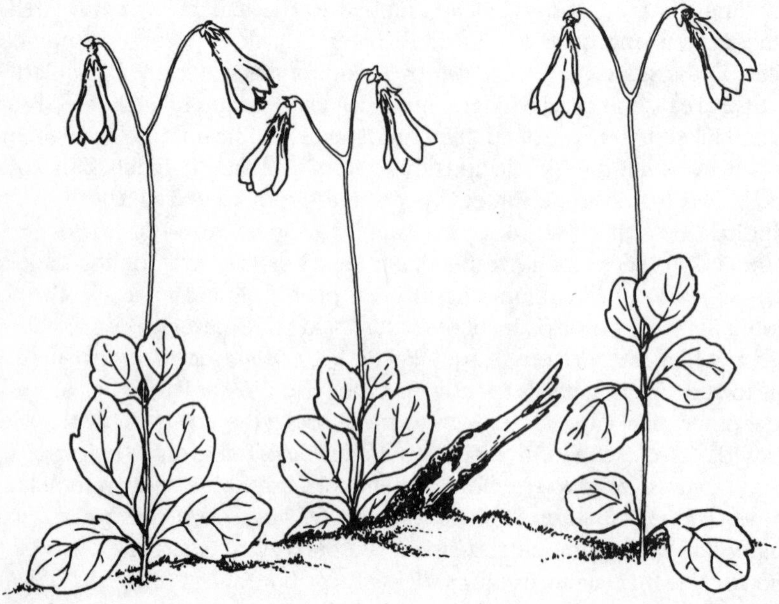

TWINFLOWER
Linnaea borealis

unifoliata), Twinflower *(Linnaea borealis),* and a particularly attractive Bellflower, *Campanula scouleri.* The Columbines are represented by a single species, the fiery red *Aquilegia formosa.*

Mount Rainier National Park is located southeast of Tacoma. The southern entrances can be reached via S.R. 706 or S.R. 123, and those on the eastern side via S.R. 410.

THE OLYMPIC PENINSULA

Separated from Canada by the wide Strait of Juan de Fuca and nearly severed from the rest of Washington by Puget Sound and a number of other waterways, the Olympic Peninsula forms the extreme northwestern corner of the contiguous United States. It was reputedly discovered in 1592 by de Fuca, a Greek explorer sailing under the flag of Spain.

Evidence of the federal government's long-standing interest in the peninsula is everywhere. At the heart of the peninsula are the Olympic Mountains, most of them now within either a national park (which extends its protection over the luxuriant rain forests as well) or a national forest, and along its wild Pacific coastline is a string of Indian reservations and national wildlife refuges.

It is possible to encircle the peninsula by taking U.S. 101 and connecting side roads (there are no highways crossing the interior). The trip can be taken in either direction, but a better sense of contrast with modern, busy Seattle will be had by ferrying across Puget Sound and driving around counterclockwise.

Approaching in this way, you are immediately struck by the parched appearance of the landscape. The explanation comes to you quickly, though: This is the corollary to the lush, dripping rain forests you will see farther west. Here you are on the lee side of the Olympic range—in its "rain shadow"—and there is very little moisture left in the clouds by the time they arrive here. At Sequim, for example, the average rainfall is only 16 inches a year.

Olympic National Park

The Olympic Mountains, which occupy the northern part of the peninsula, are the product of continental drift. Geologically they are very young, and their sharp contours have not yet been softened by erosion. Although they are not high by western standards, it was not until 1890 that the first climbers were able to penetrate the peninsula and ascend the tallest peak, 7,965-foot Mount Olympus.

Olympic National Park owes its beginnings to President Theodore Roosevelt, who established a refuge here in 1909 to protect the indigenous wapiti (since named Roosevelt elk in his honor) from being hunted to extinction.

GLACIER LILY
Erythronium grandiflorum

At one time the Olympic Mountains were isolated from the Cascades by a gigantic sheet of glacial ice, which served to bar many animal species native to the latter from migrating westward. Among them were a number of potentially destructive herbivores, and their exclusion greatly benefited the development of an alpine flora on what eventually became the Olympic peninsula. Unfortunately, however, mountain goats were introduced in the 1920s, and their progeny now pose a serious threat to the mountain vegetation.

The flower-strewn meadows for which the Olympics are noted have by no means been destroyed, however, and one of the best ways to experience them is to drive south from Port Angeles on the 18-mile paved park road to Hurricane Ridge, where you will find a lodge, picnic areas, and trails.

In its climb of nearly a mile from sea level through the Canadian zone into the Hudsonian, the road passes through a succession of evergreen tree species similar to those found on the slopes of Mount Rainier. In the upper part of the subalpine zone, those trees that have been able to survive the wind and snow are reduced to twisted *krummholz*. Farther down, where groves of erect conifers are scattered across the meadows, another phenomenon occurs. The lowest branches of several species—but notably Subalpine Fir—are pressed down into contact with the soil by the weight of the snow and sometimes take root; this layering creates a "skirt" of young trees around the base of the parent.

Near the lodge is Big Meadow, green and wet and studded with wildflowers. Three miles to the west lies Hurricane Hill, the approach to which begins with a 1.5-mile extension of the road, followed by a trail of the same length, also paved. For a third of its length the trail is comparatively level and has interpretive signs. The displays of wildflowers here are exceptional, but you should also check the beginning of the Little River Trail (which branches off to the right) and the Hurricane Hill Nature Trail loop at the summit.

If you are experienced in mountain driving, you may also want to try the road that follows the crest of the ridge in the opposite direction to Obstruction Point. This is a narrow 8-mile dirt road without guard rails that ascends to an altitude of 6,450 feet.

The wildflower season on Hurricane Ridge lasts from June to September, but normally the peak is during the first half of July. The very earliest ones burst into bloom as soon as the snow patches begin to melt. Glacier and Avalanche Lilies *(Erythronium grandiflorum* and *E. montanum),* the handsome Western Marsh Marigolds *(Caltha leptosepala)* and the large-flowered Subalpine Buttercups *(Ranunculus eschscholtzii)* form carpets of white and gold. Later, mats of Cliff Douglasia *(Douglasia laevigata)* bear vivid pink blossoms. The ensuing parade includes scores of species, among them such showy examples as Magenta Paintbrush, Subalpine Lupine, American Bistort, and Silky Phacelia *(Phacelia sericea),* as well as Partridge Foot, Narrow-petaled Stonecrop *(Sedum stenopetalum),* and Olympic Onion *(Allium crenulatum).*

An interesting feature of the Olympic Mountains flora is its high rate of endemism. One of its most charming wildflowers, which grows nowhere else in the world, is Piper's Harebell *(Campanula piperi).* These little plants with hollylike leaves and appealing blue flowers can be found atop Hurricane Ridge, where they huddle in rock crevices. Other endemics to be looked for in similar habitats are the purple-flowered Olympic Violet *(Viola flettii)* and the yellow Olympic Butterweed *(Senecio neowebsteri).*

PIPER'S HAREBELL
Campanula piperi

The Northwest Coast

In making the swing around the peninsula, one should not fail to turn outward occasionally to see what the coastlines have to offer. A major attraction along the north shore are the tidepools, depressions in the rocky ledges at the base of the cliffs, where at low tide you can examine at close range some of the sea's most fascinating flora and fauna. Plants are represented by seaweeds, species of algae, in an infinite variety of shapes and colors. Most of the animals have locked themselves into place against the turbulence or move about very slowly. There are sea urchins and starfish in rich shades of yellow, red, or purple, and many kinds of crabs, barnacles, and snails. Perhaps the most beautiful of all are the flowerlike sea anemones, which wave their fringe of tentacles when feeding. Their color is an exquisite aquamarine, derived from a species of alga living within the creature's body. One popular site is at Tongue Point, 12 miles west of Port Angeles and 4 miles north of S.R. 112, where there is a public park.

For contrast with the delicacy of these organisms, walk down to the ocean beach on the western side of the peninsula. Here the Pacific is seen not as the placid expanse that its name would imply but as an instrument of awesome power gathered on its unimpeded 5,000-mile drive all the way from Japan.

Even in its quieter moments strong surf crashes against the shore, rearranging the wreckage of previous assaults to be flung against the land by the next winter's storms. Enormous trees that once grew close to the edge had the ground cut out from under them and toppled into the sea; now reduced to whitened boles, they lie strewn about like gargantuan jackstraws. Rock fragments, tumbled and ground until they look like giant pebbles, are piled up on the shore in uncountable numbers. And sand—that innocent-appearing but relentlessly efficient tool of attrition—awaits in crescent-shaped beaches.

Nowhere is the savagery of the sea more evident than where it has attacked old headlands, chopping away at the rock until all that remains are isolated "sea stacks" standing offshore. When they are thinly veiled by fog, the imagination transforms their gaunt silhouettes into fleets of ghostly, battered ships.

The scenery near the Indian fishing village of La Push is especially dramatic, and it is well worth taking the 14-mile side road from U.S. 101. Farther south, an elevated view of the ocean can be had from the cliff tops at Kalaloch, where the main highway meets the coast. This is a good vantage point for watching the gray whales during their spring and fall migrations.

The Pacific Rain Forests

In the tropics rain forests are abundant and extensive, but in temperate climates they occur only in very special places, and the Olympic Peninsula has the finest in the world; the best known, each named for the river that winds through its glacier-carved valley to the sea, are Hoh, Queets, and Quinault.

Although both types share the name "rain forest," there are more differences than similarities. In the tropics, the substrate is poor with only a thin veneer of nutritious soil, but here the humus lies rich and deep. Instead of broad-leaved evergreens, climbing lianas, and large epiphytes, these cool forests are dominated by tall conifers, supplemented by broad-leaved deciduous trees.

Just as aridity is the basis for defining a desert, copious rainfall is necessary to qualify a rain forest, and those on the Olympic Peninsula have no trouble meeting this criterion. Warm, moisture-laden winds blowing off the Pacific are deflected by the mountains and, cooling when they reach higher altitudes, discharge massive amounts of precipitation. Much of this is in the form of rain, sometimes amounting to 200 inches in a single year, but in the upper elevations it falls as snow to perpetually replenish the snowfields and glaciers, whose meltwater provides a year-round supplement to the forests lower down. In summer, which might otherwise be a "dry season," moisture rolls in from the sea as fog.

The atmosphere in a temperate rain forest is unique. Wetness is everywhere, but it has the sweet freshness of life rather than the dank sour smell of decay. Trees soar overhead to shut out much of the sky, allowing only softly filtered sunlight to dapple the mossy ground carpet and pick out an occasional flower. Spreading branches are festooned with mosses and lichens and decorated with sprays of ferns. The entire scene is imbued with an emerald translucence and enveloped in a silence so pervasive that one cannot imagine making a harsh or discordant sound.

If there is time to see only one example, the best choice would be the Hoh River Rain Forest. It is easily accessible by a 19-mile paved road running east from U.S. 101, and with a visitor center, three self-guiding loop nature trails, and ranger-led walks, it is decidedly education oriented. Then there is always the option of getting away from the inevitable crowds by taking a trail upstream beside the Hoh River. This is the customary route to Mount Olympus, and if you were to go far enough it would put you at Blue Glacier. Anywhere here you might expect to see evidence of Roosevelt elk, black-tailed deer, or black bear.

The Hoh Forest not only receives more than 140 inches of rain each year but is fed by the runoff from the longest glacier on Mount Olympus, where annual snowfall may total 200 *feet.* Superb examples of all the major rain forest tree species may be seen here. One Sitka Spruce is over 250 feet high, and there are enormous specimens of Western Hemlock, Western Red Cedar, and Douglas Fir. Among the hardwoods, the largest known Red Alder *(Alnus rubra)* was found in the Hoh Valley.

Sitka Spruce *(Picea sitchensis),* by far the largest American species of its genus, is confined to the Pacific coastal areas of the United States and Canada. It can be recognized by its whitened needles, which are unusual for a spruce in that they are flat in crosssection. During World War I the wood of Sitka Spruce was used extensively in airplane construction.

Except for its huge size, Western Hemlock *(Tsuga heterophylla)* might be taken for its eastern counterpart, *T. canadensis,* even to the gracefully drooping tip. Completely adapted to growing in the shade, it is the closest thing to a rain forest climax species. Similarly, a Western Red Cedar *(Thuja plicata)* is much like a Northern White Cedar *(T. occidentalis)* enlarged three or four times, except that its cones still do not exceed one-half inch in length. It is this tree that furnished the durable wood from which the coastal Indians made their war canoes and totem poles.

Douglas Fir *(Pseudotsuga menziesii)* is a splendid tree, and can exceed in height all other trees except the *Sequoias.* Obviously it is not a true fir, for the cones are pendent and fall from the trees intact. The long three-pointed bracts that protrude from between the scales are diagnostic.

Since water is so abundant near the surface there is no need for trees to send their roots deep into the ground for moisture, and with only shallow support in the spongy, sodden soil they frequently come crashing down. As they lie moldering these logs acquire a covering of moss, which retains water and collects detritus. In this way they become the rain forest's primary seedbed, in which seeds from other trees can germinate, and in time these fallen trunks become "nurse logs" with incipient forest giants arising from them in rows, their roots reaching down along the sides to find the soil. Although many kinds of

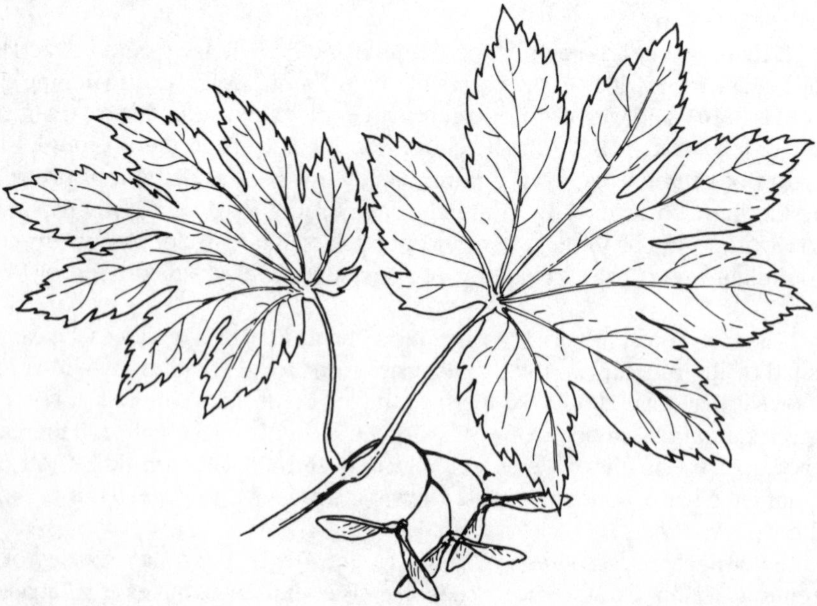

VINE MAPLE
Acer circinatum

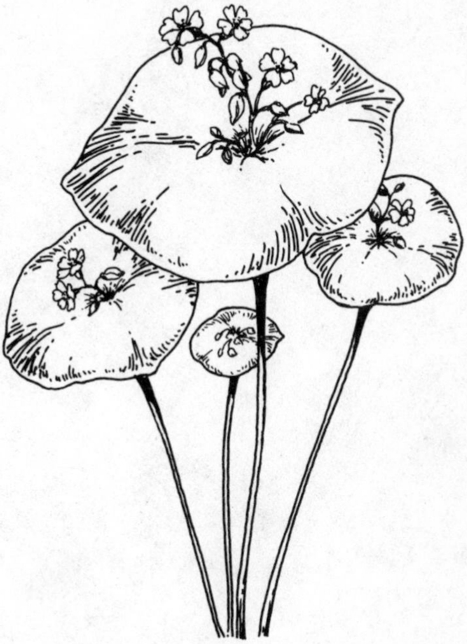

MINER'S LETTUCE
Montia perfoliata

trees, as well as shrubs, ferns, and herbaceous species, get their start in this way, the ones most likely to attain optimum size are the shade-tolerant Western Hemlock and Sitka Spruce.

A large and conspicuous tree of the middle story is Bigleaf Maple *(Acer macrophyllum).* Its leaves, which may be a foot in length, are an apparent adaptation for light gathering in the dim forest. A very different species, *A. circinatum,* or Vine Maple, usually grows as a tall shrub, and often forms thickets as its arching branches take root on contact with the ground. It has leaves that are round in outline and, unlike other native maples, have seven to nine lobes. Both of these maples are frequently joined along stream banks by Red Alder and Black Cottonwood *(Populus trichocarpa).*

The wet environment is ideal for the more primitive plant forms. Species of mosses run into the scores, and there are fungi, lichens, liverworts, spike mosses, and slime molds. The showiest are, of course, the ferns, and prominent among them are Sword Fern *(Polystichum munitum),* which is the commonest; Lady Fern *(Athyrium filix-femina);* and Maidenhair Fern *(Adiantum pedatum* var. *aleuticum).* The sweet-flavored rootstocks of Licorice Fern *(Polypodium glycorrhiza)* grow on moss-covered tree branches.

Because of the dense shade, wildflower species are relatively few, but some worthy of mention are Deerfoot Vanilla Leaf *(Achlys triphylla),* Redwood Sorrel *(Oxalis oregana),* Miner's Lettuce *(Montia perfoliata),* Beadruby *(Maianthemum dilatatum),* and the yellow Pioneer's Violet *(Viola glabella).*

APPENDIX

Botanic Gardens and Arboretums

ALABAMA

Bellingrath Gardens, Theodore, AL (20 mi. south of Mobile)

Birmingham Botanical Garden, 2612 Lane Park Road, Birmingham, AL

University of Alabama Arboretum, Tuscaloosa, AL

William Bartram Arboretum, Wetumpka, AL (in Fort Toulouse-Jackson Park)

ARIZONA

Boyce Thompson Southwestern Arboretum, 3 mi. west of Superior, AZ

The Arboretum at Flagstaff, Woody Mountain Road, Flagstaff, AZ

Desert Botanical Garden, 1201 North Galvin Parkway, Phoenix, AZ (in Papago Park)

CALIFORNIA

Berkeley Botanical Garden, Centennial Drive, Berkeley, CA (adjacent to University of California main campus)

Huntington Botanical Gardens, 1151 Oxford Road, San Marino, CA

The Living Desert, 47-900 Portola Avenue, Palm Desert, CA

Mendocino Coast Botanical Gardens, Fort Bragg, CA

Mildred E. Mathias Botanical Garden, Hilgard and Le Conte Avenues, Los Angeles, CA

Moorten Botanical Garden, 1701 South Palm Canyon Drive, Palm Springs, CA

Quail Botanical Gardens, 230 Quail Garden Drive, Encinitas, CA

Rancho Santa Ana Botanic Garden, 1500 North College Avenue, Claremont, CA

Santa Barbara Botanic Garden, 1212 Mission Canyon Road, Santa Barbara, CA

Charles Lee Tilden Regional Park Botanical Garden, Berkeley, CA

University of California Arboretum, Davis, CA (on campus)

University of California at Riverside Botanic Gardens, Riverside, CA (on east side of campus)

COLORADO

Chatfield Arboretum of Denver Botanic Gardens, Denver, CO (near Chatfield Dam southwest of Denver)

Mount Goliath Alpine Unit of Denver Botanic Gardens, 50 mi. west of Denver, CO

CONNECTICUT

Connecticut College Arboretum, Connecticut College (William Street Gate), New London, CT

Audubon Fairchild Garden, North Porchuck Road, Greenwich, CT

White Memorial Conservation Center, U.S. 202, Litchfield, CT

DELAWARE

Winterthur Museum and Gardens, Winterthur, DE

DISTRICT OF COLUMBIA

United States National Arboretum, 3501 New York Avenue, N.E., Washington, DC

United States Botanic Garden, Maryland Avenue, S.W., Washington, DC (at foot of Capitol Hill)

FLORIDA

Bok Tower Gardens, Mountain Lake Sanctuary, Lake Wales, FL

Fairchild Tropical Garden, 10901 Old Cutler Road, Coral Gables, FL

Joe Allen Garden Center, West Martello Tower, Key West, FL

Marie Selby Botanical Gardens, 800 South Palm Avenue, Sarasota, FL

Ravine State Gardens, Palatka, FL

Washington Oaks State Gardens, Route 1, St. Augustine, FL

Thomas Edison Winter Home and Botanical Gardens, 2350 McGregor Boulevard, Fort Myers, FL

GEORGIA

Callaway Gardens, U.S. 27, Pine Mountain, GA (75 mi. southwest of Atlanta)

Fernbank Greenhouse and Botanical Gardens, 765 Clifton Road, N.E., Atlanta, GA

State Botanical Garden of Georgia, 2450 South Milledge Avenue, Athens, GA

IDAHO

Charles Huston Shattuck Arboretum, University of Idaho, Moscow, ID

ILLINOIS

Abraham Lincoln Memorial Garden, Lake Springfield Park, Springfield, IL

Chicago Botanic Garden, Lake Cook Road, Chicago, IL (22 mi. north of Chicago Loop)

Center for Natural Landscaping, 2603 Sheridan, Evanston, IL

Early American Museum and Botanical Gardens, Lake of the Woods Park, Mahomet, IL

Morton Arboretum, Route 53 north of East-West Tollway, Lisle, IL (25 mi. west of downtown Chicago)

INDIANA

Hayes Regional Arboretum, 801 Elks Road, Richmond, IN

Huntington College Arboretum and Botanical Garden, 2303 College Avenue, Huntington, IN

Jerry E. Clegg Botanical Garden, 1854 North 400th East, Lafayette, IN

IOWA

Bickelhaupt Arboretum, 340 South 14th Street, Clinton, IA

Biological Preserves System, University of Northern Iowa, Cedar Falls, IA

KANSAS

Dyck Arboretum of the Plains, Hesston, KS

Kansas Landscape Arboretum, Milford Lake, Wakefield, KS

KENTUCKY

Bernheim Forest Arboretum, S.R. 245, Clermont, KY (30 mi. south of Louisville)

LOUISIANA

Audubon Park and Zoological Garden, between Magazine Street and St. Charles Avenue, New Orleans, LA

Hilltop Arboretum, 11800 Highland Road, Baton Rouge, LA

Hodges Garden, S.R. 171, Many, LA (between Shreveport and Lake Charles)

Louisiana Purchase Arboretum, Chicot State Park, Ville Platte, LA

MAINE

Fay Hyland Botanical Plantation, University of Maine, Orono, ME

Perkins Arboretum and Bird Sanctuary, Colby College, Waterville, ME

MARYLAND

Brookside Gardens, 1500 Glenallan Avenue, Wheaton, MD

Cylburn Wildflower Preserve, 4915 Greenspring Avenue, Baltimore, MD

MASSACHUSETTS

Arnold Arboretum, Arborway, Jamaica Plain, Boston, MA

Botanic Garden of Smith College, West and Elm Streets, Northampton, MA

Botanic Trails Historical Society, Gatehouse on Kingshighway, Yarmouth Port, MA

Garden in the Woods, New England Wild Flower Society, Hemenway Road, Framingham, MA

Norcross Wildlife Sanctuary, Monsom-Wales Road, Wales, MA

MICHIGAN

Cranbrook Gardens, 380 Lone Pine Road, Bloomfield Hills, MI

Fernwood Botanic Garden and Nature Center, 1720 Range Line Road, Niles, MI

For-Mar Nature Preserve and Arboretum, 5360 East Potter Road, Flint, MI

Hidden Lake Gardens, Michigan State University, Tipton, MI

Dow Gardens, 1018 West Main Street, Midland, MI

Matthaei Botanical Garden, University of Michigan, 1800 North Dixboro Road, Ann Arbor, MI

MINNESOTA

Minnesota Landscape Arboretum, 3675 Arboretum Drive, Chanhassen, MN

Eloise Butler Wildflower and Bird Sanctuary, 3800 Bryant Avenue South, Minneapolis, MN

MISSISSIPPI

Grey Oaks Gardens, 4142 Rifle Range Road, Vicksburg, MS

MISSOURI

Missouri Botanical Gardens, 2101 Tower Grove Avenue, St. Louis, MO

Shaw Arboretum, I-44 and S.R. 100, Gray Summit, MO (35 mi. southwest of St. Louis)

NEVADA

Botanical Garden, Lake Mead National Recreation Area, Boulder City, NV

New Hampshire

Lost River Nature Garden, Kinsman Notch, NH

New Jersey

Cora Hartshorn Arboretum and Bird Sanctuary, 324 Forest Drive South, Short Hills, NJ

Pack Memorial Arboretum, Washington Crossing State Park, 8 mi. northwest of Trenton on S.R. 29

Skyland Gardens, Ringwood State Park, Ringwood, NJ

New Mexico

Living Desert State Park, U.S. 285, Carlsbad, NM

New York

Bayard Cutting Arboretum, Montauk Highway, Great River, NY

Brooklyn Botanic Garden, 1000 Washington Avenue, Brooklyn, NY

Cornell Plantations, Cornell University, 1 Plantation Road, Ithaca, NY

New York Botanical Garden, Bronx Park, Bronx, NY

Old Westbury Gardens, 71 Old Westbury Road, Old Westbury, NY

Planting Fields Arboretum, Planting Fields Road, Oyster Bay, NY

North Carolina

Biltmore House and Gardens, U.S. 25 off I-40, Asheville, NC

Highlands Nature Center and Botanical Garden, Highlands, NC

North Carolina Botanical Garden and Coker Arboretum, Laurel Hill Road, Chapel Hill, NC

Pearson's Falls, off N.C. 176 between Tryon and Saluda, NC

Sarah P. Duke Memorial Gardens, Duke University, Durham, NC

University Botanical Gardens, 151 W. T. Weaver Boulevard, Asheville, NC

Ohio

Adell Durban Park and Arboretum, Stow, OH

Dawes Arboretum, 7770 Jacksontown Road, Newark, OH

George P. Crosby Gardens, 5403 Elmer Drive, Toledo, OH

Holden Arboretum, 9500 Sperry Road, Mentor, OH

Secor Metropark, Oak Openings Parkway, Toledo, OH

Oklahoma

Forest Heritage Center, Beavers Bend State Park, Broken Bow, OK

Oregon

Castle Crest Wildflower Garden, Crater Lake, OR

Hoyt Arboretum, 4000 Fairview Boulevard, Portland, OR

Mount Pisgah Arboretum, Mount Pisgah Road, Eugene, OR

Pennsylvania

Bartram's Garden, 54th Street and Lindbergh Boulevard, Philadelphia, PA

Bowman's Hill Wildflower Preserve, Washington Crossing Historical State Park, Washington Crossing, PA (south of New Hope)

Jenkins Arboretum, 631 Berwyn Road, Devon, PA

Longwood Gardens, Kennett Square, PA

Rhode Island

Blithewold Gardens and Arboretum, Ferry Road, Bristol, RI

South Carolina

Brookgreen Gardens, U.S. 17, Murrells Inlet, SC
Clemson University Horticultural Gardens, Clemson, SC
Cypress Gardens, Charleston, SC (25 mi. north off S.R. 52)
Kalmia Gardens, Coker College, Hartsville, SC
Magnolia Plantation and Gardens, S.R. 61, Charleston, SC
Middleton Place, S.R. 61, Charleston, SC

Tennessee

Memphis Botanic Garden, 750 Cherry Road, Memphis, TN (in Audubon Park)
Rock City Gardens, 1400 Patten Road, Lookout Mountain, Chattanooga, TN
Tennessee Botanical Gardens and Fine Arts Center at Cheekwood, Forrest Park Drive, Nashville, TN (8 mi. southwest of downtown)

Texas

Arboretum Incorporated, Route 1, Arp, TX
Brazos County Arboretum, Texas A & M University, College Station, TX
Houston Arboretum and Botanical Gardens, 4500 Woodway Drive, Houston, TX
San Antonio Botanical Center, 555 Funston Place, San Antonio, TX

Utah

Brigham Young University Arboretum, Provo, UT

Vermont

Stone Chimney Garden, Reading, VT
Vermont Wildflower Farm, Charlotte, VT

Virginia

Monticello, S.R. 53, Charlottesville, VA
Norfolk Botanical Gardens, Airport Road, Norfolk, VA

Washington

Ohme Gardens, 3327 Ohme Road, Wenatchee, WA
Sehome Mill Arboretum, Western Washington University, Bellingham, WA
University of Washington Arboretum, Washington Park, Seattle, WA

West Virginia

Brooks Memorial Arboretum, Watoga State Park, Caldwell, WV
Core Arboretum, West Virginia University, Morgantown, WV

Wisconsin

Jones Arboretum, U.S. 14, Readstown, WI
Mackenzie Environmental Education Center, Poynette, WI
Putnam Park, University of Wisconsin, Eau Claire, WI
Boerner Botanical Gardens, Hales Corners, WI (in Whitnall Park)
University of Wisconsin Arboretum, 1207 Seminole Highway, Madison, WI

Index

Common names of plants are set in roman type, scientific names in *italics*. Pages on which illustrations appear are indicated by **boldface** type.